Lecture Notes in Earth Sciences

Edited by Somdev Bhattac
Horst J. Neugebauer and A

21

Ulrich Förstner

Contaminated Sediments

Lectures on Environmental Aspects
of Particle-Associated Chemicals in Aquatic Systems

Springer-Verlag
Berlin Heidelberg New York London Paris Tokyo

Author

Prof. Dr. Ulrich Förstner
Arbeitsbereich Umweltschutztechnik
Technische Universität Hamburg-Harburg
Eissendorferstr. 40, D-2100 Hamburg 90, FRG

ISBN 3-540-51076-1 Springer-Verlag Berlin Heidelberg New York
ISBN 0-387-51076-1 Springer-Verlag New York Berlin Heidelberg

Printed in Germany

Printing and binding: Druckhaus Beltz, Hemsbach/Bergstr.
2132/3140-543210 – Printed on acid-free paper

To

Professor Imig

and Uta

for helping me to recover

TABLE OF CONTENTS

1. INTRODUCTION

Sediments are increasingly recognized as both a **carrier** and a possible **source of contaminants** in aquatic systems, and these materials may also affect groundwater quality and agricultural products when disposed on land. Contaminants are not necessarily fixed permanently by the sediment, but may be recycled via biological and chemical agents both within the sedimentary compartment and the water column. Bioaccumulation and food chain transfer may be strongly affected by sediment-associated proportions of pollutants. Benthic organisms, in particular, have direct contact with sediment, and the contaminant level in the sediment may have greater impact on their survival than do aqueous concentrations. Following the findings of positive correlations between liver lesions in English Sole and concentrations of certain aromatic hydrocarbons in Puget Sound (Washington) sediment, it can be suspected that such substrates may also be responsible for a host of other serious and presently unrecognized changes at both the organismal and ecosystem levels (Malins *et al.*, 1984).

Modern research on particle-bound contaminants probably originated with the idea that sediments reflect the biological, chemical and physical conditions in a water body (Züllig, 1956). Based on this concept the **historical evolution** of limnological parameters could be traced back from the study of vertical sediment profiles. In fact, already early in this century Nipkow (1920) suggested that the alternative sequence of layers in a sediment core from Lake Zürich might be related to variations in the trophic status of the lake system. During the following decades of limnological research on **eutrophication problems** sediment aspects were playing only a marginal role, until it was recognized that recycling from bottom deposits can be a significant factor in the nutrient budget of an aquatic system. Similarly, in the next global environmental issue, the **acidification** of inland waters sediment-related research only became gradually involved. Here too, it is now accepted that particle-interactions can affect aquatic ecosystems, e.g. by enhancing the mobility of toxic metals.

In contrast to the eutrophication and acidification problems, research on **toxic chemicals** has included sediments aspects from its beginning: Artificial radionuclides in the Columbia and Clinch Rivers in the early sixties (Sayre *et al.*, 1963); in the late sixties heavy metals in the Rhine River system (De Groot, 1966) and methyl mercury (Jensen & Jernelöv, 1967) at Minamata Bay in Japan, in Swedish lakes, in Alpine Lakes, Laurentian Great Lakes and in the Wabigoon River system in Canada; organochlorine insecticides and PCBs in Lakes St. Clair and Erie during the seventies (Frank *et al.*, 1977); chlorobenzenes and TCDDs in the Niagara River system and Lake Ontario in the early eighties (Oliver & Nicol, 1982; Smith *et al.*, 1983).

In the present lecture notes, following the description of priority pollutants related to sedimentary phases (Chapter 2), four aspects will be covered, which in an overlapping succession also reflect the development of knowledge in particle-associated pollutants during the past twenty-five years:

- the identification, surveillance, monitoring and control of **sources** and distribution of pollutants (Chapter 3);
- the evaluation of **solid/solution relations** of contaminants in surface waters (Chapter 4);
- the study of **in-situ processes** and mechanisms in pollutant transfer in various compartments of the aquatic ecosystems (Chapter 5);
- The assessment of the **environmental impact** of particle-bound contaminants, i.e. the development of sediment quality criteria (Chapter 6).

A final chapter will focus on practical aspects with contaminated sediments. Available technologies will be described as well as future perspectives for the management of **dredged materials**. Here too, validity of remedial measures can only be assessed by integrated, multidisciplinary research.

In the view of the growing information on the present subject and owing to the limitations in the framework of this monography, the reader is referred to additional **selected bibliography**, which is attached at the end of this Chapter 1. Additional information on the more recent publications on contaminated sediments is given in the annual review volume of the Journal of the Water Pollution Control Federation, June edition.

References

De Groot, A.J. (1966) Mobility of trace metals in deltas. In *Meeting Int. Comm. Soil Sciences, Aberdeen*, ed. G.V. Jacks, Trans. Comm. II & IV, pp. 267-97.

Frank, R. et al. (1977) Organochlorine insecticides and PCBs in sediments of Lake St. Clair (1970 and 1974) and Lake Erie. *Sci. Tot. Environ.* 8, 205-27.

Jensen, S. & Jernelöv, A. (1967) Biosynthesis of methyl mercury. *Nordforsk Biocid Inform.* (Swedish) 10, 4-12.

Malins, D.C. *et al.* (1984) Chemical pollutants in sediments and diseases of bottom-dwelling fish in Puget Sound, Washington. *Environ. Sci. Technol.* 18, 705-713.

Nipkow, F. (1920) Vorläufige Mitteilungen über Untersuchungen des Schlammabsatzes im Zürichsee. *Z. Hydrol.* 1, 101-23.

Oliver, B.G. & Nicol, K.D. (1982) Chlorobenzenes in sediments, water, and selected fish from Lakes Superior, Huron, Erie, and Ontario. *Environ. Sci. Technol.* 16, 532-536.

Sayre, W.W., Guy, H.P. & Chamberlain, A.R. (1963) Uptake and transport of radionuclides by stream sediments. *U.S. Geol. Surv. Prof. Pap.* 433-A, 23 p.

Smith, R.M. *et al.* (1983) 2,3,7,8-tetrachlorodibenzo-*p*-dioxin in sediment samples from Love Canal storm sewers and creeks. *Environ. Sci. Technol.* 17, 6-10.

Züllig, H. (1956) Sedimente als Ausdruck des Zustandes eines Gewässers. *Schweiz. Z. Hydrol.* 18, 7-143.

Books for Further Reading

Allan, R.J. (1986) *The Role of Particulate Matter in the Fate of Contaminants in Aquatic Ecosystems*. National Water Research Institute, Scientific Series No. 142. 128 p. Burlington/Ontario: Canada Centre for Inland Waters.

Allen, H.E. (Ed.)(1984) *Micropollutants in River Sediments*. Report on a WHO Working Group Meeting, Trier/FRG, August 5-8, 1980. EURO Reports and Studies No. 61, 85 p. Copenhagen: WHO Regional Office for Europe.

Baker, R.A. (Ed.)(1980) *Contaminants and Sediments*. 2 Vols. Ann Arbor, Michigan: Ann Arbor Sci. Publ.

Dickson, L., Maki, A.W. & Brungs, W.A. (Eds.)(1987) *Fate and Effects of Sediment-Bound Chemicals in Aquatic Systems*. 449 p. New York: Pergamon Press.

Goldberg, E.D. (Ed.)(1978) *Biogeochemistry of Estuarine Sediments.* 293 p. Paris: UNESCO.

Golterman, H.L. (Ed.) (1977) *Interactions between Sediments and Fresh Water.* Proc. 1st Int. Symp. Amsterdam, Sept. 6-10, 1976. 472 p. The Hague and Wageningen: Junk and PUDOC.

Golterman, H.L., Sly, P.G. & Thomas, R.L. (Ed.)(1983) *Study of the Relationship between Water Quality and Sediment Transport. A Guide for the Collection and Interpretation of Sediment Quality Data.* 231 p. Paris: UNESCO Press.

Hart, B.T. & Sly, P.G. (Eds.)(1989) *Sediments and Freshwater Interactions.* Proc. 4th Intern. Symp., Melbourne, Febr. 17-20, 1987. The Hague: Dr. W. Junk, Publ. (in press).

Kavanaugh, M.C. & Leckie, J.O. (Eds.)(1980) *Particulates in Water - Characterization, Fate, Effects, and Removal.* 401 p. Washington, D.C.: American Chemical Society. (*Advances in Chemistry Ser.* 189)

McCall, P.L. & Tevesz, M.J.S. (Eds.)(1986) *Animal-Sediment Relations - The Biogenic Alteration of Sediments.* New York: Plenum Press.

Postma, H. (Ed.)(1981) *Sediment and Pollution Exchange in Shallow Seas.* Rapp. P.-v. Réun. Cons. Intern. Explor. Mer (ICES), Vol. 181, 223 p. Copenhagen: ICES.

Salomons, W. & Förstner, U. (Eds.)(1988) *Chemistry and Biology of Solid Waste - Dredged Material and Mine Tailings.* 305 p. Berlin: Springer-Verlag.

Shear, H. & Watson, A.E.P. (Eds.)(1977) *The Fluvial Transport of Sediment-Associated Nutrients and Contaminants.* Proc. Int. Workshop Kitchener/Ontario, Oct. 20-22, 1976. 309 p. Windsor/Ontario: International Joint Commission.

Sly, P. (Ed.(1982) *Sediment/Freshwater Interactions.* Proc. 2nd Intern. Symp. Kingston, Ontario, June 15-18, 1981. 701 p. The Hague: Dr. W. Junk Publ. (*Hydrobiologia* Vols. 91/92).

Sly, P. (Ed.)(1986) *Sediment and Water Interactions.* Proc. 3rd Intern. Symp. Geneva/Switzerland, August 27-31, 1984. 521 p. New York: Springer-Verlag.

Thomas, R.L. *et al.* (Eds.)(1987) *Ecological Effects of In-Situ Sediment Contaminants.* Proc. Int. Workshop Aberystwyth, Wales, August 21-24, 1984. 272 p. Dordrecht/The Netherlands: Dr. W. Junk Publ. (*Hydrobiologia*, Vol. 149)

2. PRIORITY POLLUTANTS

2.1. Historical Development

Geochemical investigations of **stream sediment** have long been standard practice in **mineral exploration** (Hawkes & Webb, 1962); by more extensive sampling and analysis of metal contents in water, soils and plants, the presumable enrichment zones can be narrowed down and, in favourable cases, localized as exploitable deposits. Generally, the variation in trace metal content of stream sediments can be characterized as function of potential controlling factors "influence of lithologic units", "hydrologic effects", "geologic features", "cultural (man-made) influences", "type of vegetational cover", and "effects of mineralized zones" (Dahlberg, 1968). Similarly, **lake sediment geochemistry** has been used as a guide to mineralization, particularly intensive on lakes of the Canadian Shield (Allan, 1971). This approach attracted much attention when mineral exploration was followed by large-scale **mining and processing** activities: "Both the exploration and environmental geochemist can be looking for the same type of areas, those with high metal concentrations, but obviously from a different motivation" (Allan, 1974). It has been demonstrated that lake sediments reflect both natural processes and human activities operating within the drainage area, and changes which take place within the lake; however, "the main problem is to be able to adequately elucidate and interpret the significance of various physical and chemical attributes of sediments" (Hakanson, 1977). In this respect, the interdependence between the acceleration of **nutrient cycles** and simplification of ecosystem is typically observed as a consequence of lake pollution with organic and inorganic substances (Stumm & Baccini, 1978). Nonetheless, on a qualitative basis, sediment analysis can favourably be used to estimate **point sources** of pollutants that upon being discharged to surface waters do not remain in solution but are rapidly adsorbed by particulate matter, thereby escaping detection by water monitoring. In this sense, sediment data play an increasing role within the framework of **environmental forensic investigations** (Meiggs, 1980).

The study of dated **sediment cores** has proven particularly useful as it provides a historical record of the various influences on the aquatic system by indicating both the natural background levels and the man-induced accumulation of elements over an extended period of time. Marine and - in particular - lacustrine environments have the ideal conditions necessary for the incorporation and permanent fixing of metals and organic pollutants in sediments: reducing (anoxic) and non-turbulent environments, steady deposition, and the presence of suitable, fine-grained mineral particles for pollutant fixation. Various approaches to the **dating** of sedimentary profiles have been used but the isotopic techniques, using ^{210}Pb, ^{137}Cs and $^{239+240}Pu$, have produced the more unambiguous results and therefore have been the most successful (see review on "Historical Monitoring" by Alderton, 1985).

On the basis of data from sediment cores, Figure 2-1 (after Müller, 1981) demonstrates the development and the present situation of environmental pollution with **specific contaminants** in different parts of the world: During the last decades of the last century environmental pollution history began with an increase of **heavy metals**, reaching a maximum between about 1960-1970. These studies revealed that individual heavy metals and **polycyclic aromatic hydrocarbons** (PAH) show parallel evolution patterns and it is concluded, that both groups of pollutants could stem from a common source: combustion of coal and lignite as a consequence of increasing industrialization, chiefly in the northern hemisphere. The increasing substitution of coal and lignite by petroleum products during the past 30 years is documented by a decrease in heavy metals and PAH concentrations in the youngest sediment layers, whereas, on the other hand, **petroleum-derived hydrocarbons** show a sharp increase. The history of fecal pollution can be traced back in sediments even into pre-industrial times: **coprostanol**, one of the principal sterols of excreta of higher animals and man is an indicator of the type of pollution. Since World War II **halogenated hydrocarbons** have become significant toxic compounds in marine and other ecosystems. **Polychlorinated biphenyls**, which have been used as plasticizers in paints, plastics, resins, inks, copy paper, and adhesives (open use), first occur in sediments around 1935-1940, and although their application had been legally restricted to closed systems - dielectric fluids in transformators and capacitors, and as components of hydraulic fluids -, a decrease of concentration in the most recent sediment layers is not yet to be observed. The same holds true for **DDT** (totally banned in most western countries between 1970-1975) and its metabolites: concentrations

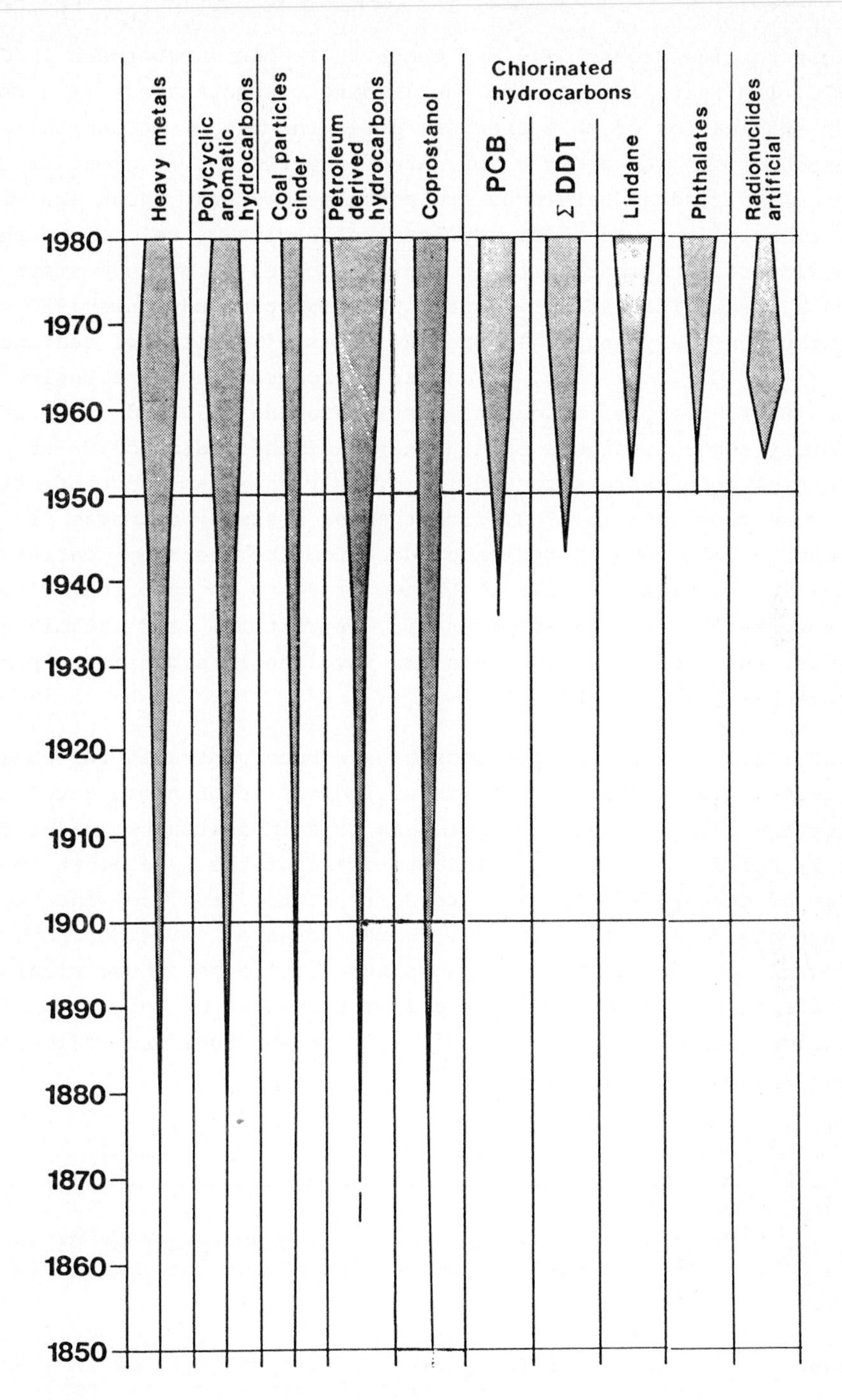

Figure 2-1 Development and Present Situation of Environmental Pollution with Specific Contaminants in Different Parts of the World as Evidenced from Dated Sediment Cores (after Müller, 1981)

begin to rise around 1945 and reach their maximum between 1960 and 1970. The sharp increase of the **lindane** concentrations is a result of the application of this chemical after the ban of "technical γ-benzene hexachloride" and other chlorinated insecticides. A group of chemicals not classified as hazardous are **phthalate esters**, which are widely used as plasticizers and their application is closely related to the development of PVC production; in sediment cores from the western Baltic Sea a steady increase of typical DEHP is observed since 1950 with a maximum in the youngest layers. A series of **artificial radionuclides** (e.g.' ^{137}Cs, $^{239,240}Pu$, ^{55}Fe) was introduced into the environment as a result of atmospheric weapons testing during 1952 and 1962. The sedimentary record reflects the intensity of the radionuclide emissions in the high atmosphere with a delay of only one year: concentrations begin to rise from 1953 to 1963, from then on a steady decrease is to be observed. (Emissions of radionuclides from the Chernobyl catastrophe in May 1986 provided a pulse which now presents an opportunity to study transport processes in atmospheric, terrestrial, and aquatic reservoirs, and specifically mechanisms involved in sedimentary processes; Santschi *et al.*, 1988).

Historical profiles of **phosphorus** have been generated for some Great Lakes by simulation from variables indicative of human development (Chapra, 1977). Figure 2-2 for Lake Erie indicates two major increases of P-loads: First during the latter part of the nineteenth century, the time of change from forested to agricultural land use, and second, since the 30's of this century, with increased sewering, population growth, and introduction of phosphate detergents. These simulations are confirmed by sediment core data from Lake Erie of both total P-concentrations and differentiation of major phosphorus forms (Figure 2-3; Williams *et al.*, 1976).

Figure 2-2: Simulation of the Historical Development of Natural and Anthropogenic Inputs of Phosphorus into Lake Erie (Chapra, 1977). *(Above)*

Figure 2-3: Profiles of Phosphorus and Its Major Forms in a Sediment Core from Western Basin of Lake Erie (Williams *et al.*, 1976). *(Below)*

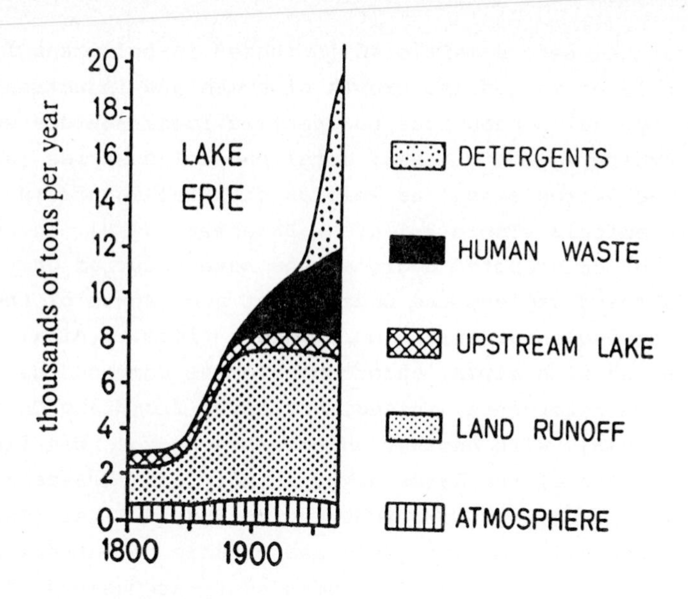

20
18
16
14
12
10
8
6
4
2
0
thousands of tons per year
1800
1900
LAKE
ERIE
DETERGENTS
HUMAN WASTE
UPSTREAM LAKE
LAND RUNOFF
ATMOSPHERE

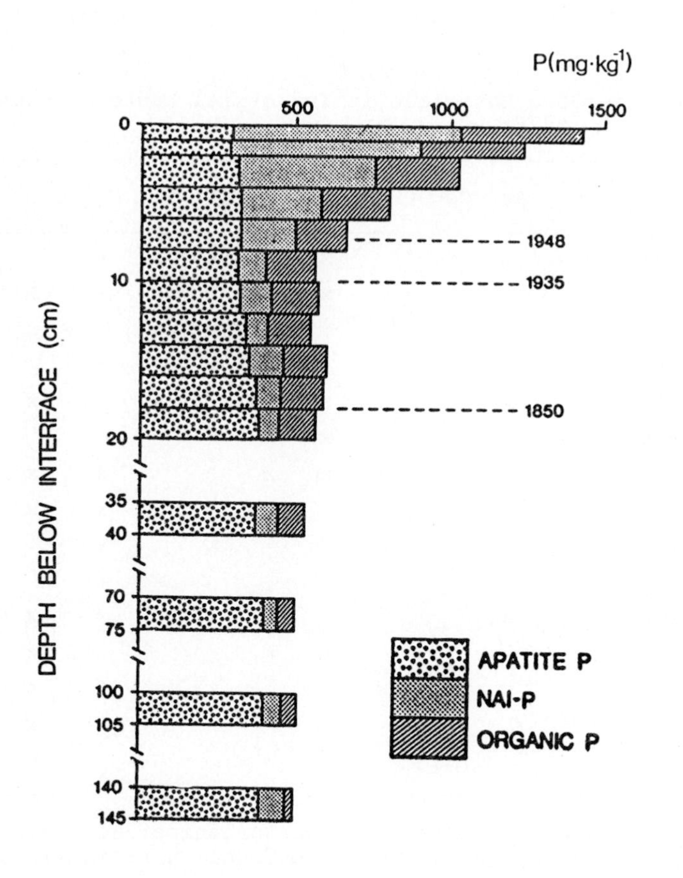

P(mg·kg⁻¹)
500
1000
1500
0
10
20
35
40
70
75
100
105
140
145
DEPTH BELOW INTERFACE (cm)
1948
1935
1850
APATITE P
NAI-P
ORGANIC P

Whereas many of the before-mentioned increases in pollutant flux to a sediment are related to general growth of human and industrial activity, there are typical connections to specific **local sources** such discharges from smelters (Cu, Ni, Pb), metal-based industries (e.g., Zn, Cr and Cd from electroplating) as well as chemical manufacturing plants (Hg, organic chemicals; Table 2-1 after Barnhart, 1978). In particular some compounds of chlorinated hydrocarbons have occurred only in the last few decades and reflect the development and growth of these more sophisticated technologies during more recent times. Typical examples have been observed with **mirex**, an organochlorine compound used in insectices and fire retardents, in sediment cores from Lake Ontario (Holdrinet *et al.*, 1978), with another chlorinated hydrocarbon insecticide, **kepone**, in sediments of the James River estuary and Chesapeake Bay (Huggett *et al.*, 1980), and with **PCBs** in the Hudson River (Turk, 1980). The historical evolution of inorganic and organic pollutants in the Niagara River will be presented as a case study at the end of this chapter.

Table 2-1 Some Hazardous Materials in Industrial Waste Streams (Barnhart, 1978)

Industry	As	Cd	CHC[a]	Cr	Cu	CN	Pb	Hg	Org[b]	Se	Zn
Mining and metallurgy	x	x		x	x	x	x	x		x	x
Paints and dyes		x		x	x	x	x	x	x	x	
Pesticides	x		x			x	x	x	x		x
Electrical & electronic			x		x	x	x	x		x	
Cleaning & duplicating	x			x	x		x		x	x	
Electroplating/finishing				x	x	x					x
Chemical manufacturing			x	x	x			x	x		
Explosives	x				x		x	x	x		
Rubber & plastics			x			x		x			x
Batteries		x					x	x			
Pharmaceuticals	x							x	x		
Textiles				x	x				x		
Petroleum & coal	x		x				x				
Pulp & paper								x	x		
Leather				x					x		

[a] chlorinates hydrocarbons [b] Miscellaneous organics: acrolein, chloropicrin, dimethyl-, dinitrobenzene, dinitrophenol, nitroaniline, etc.

2.2. Metals

Metals are natural constituents of rocks, soils, sediments, and water. However, over the 200 years following the beginning of industrialization huge changes in the **global budget** of critical chemicals at the earth's surface have occurred, challenging those regulatory systems which took millions of years to evolve (Wood & Wang, 1983). For example, the ratio of the annual mining output of a given element to its natural concentration in unpolluted soils, which can be used as an "Index of Relative Pollution Potential" (Förstner & Müller, 1973) is particularly high for Pb, Hg, Cu, Cd and Zn, namely 10 to 30 times higher than for Fe or Mn, respectively.

As for the mechanisms of **toxicity**, the most relevant is certainly the chemical inactivation of enzymes. All divalent transition metals readily react with the amino, imino and sulfhydryl groups of proteins; some of them may compete with essential elements such as zinc and displace it in metalloenzymes. Some metals may also damage cells by acting as antimetabolites, or by forming precipitates or chelates with essential metabolites. Soil biochemical processes considered especially sensitive to heavy metals are mineralization of N and P, cellulose degradation and possibly N_2-fixation. Ecotoxicological considerations in aquatic systems involve self-purification in surface and groundwater, the effects of heavy metal enrichment on biologic purification treatment, the influence on crustaceans, fish, and ultimately on man (Moore & Ramamoorthy, 1984). Elements such as silver and copper may induce adverse effects on aquatic biota far below the actual limits for drinking water. In addition to the hazard of direct toxicity to organisms, biological uptake of trace metals may lead to modification of food webs and toxicity to man through the consumption of contaminated food. With respect to the aqueous metal species it has been suggested that the "free" or aquo-metal ion form is the most available for organisms compared to the particulate, complexed or chelated forms. On the other hand, there are physical and chemical processes such as adsorption, filtration, sedimentation, complexation, precipitation, and redox reactions, which can act as partial or almost complete barriers to the movement of metals along their pathway to man (e.g., Nriagu, 1984).

The metal ions used by biologic systems must be both abundant in nature and readily available as soluble species. Abundance generally restricts the available metals to those of atomic numbers below 40, some of which are virtually unavailable due to the low solubility of their hydroxides, e.g., aluminium and titanium. Viewed from the standpoint of environmental pollution, metals may be classified according to three criteria (Table 2-2; after Wood, 1974): (i) noncritical, (ii) toxic but very insoluble or very rare, and (iii) very toxic and relatively accessible.

Table 2-2 Classification of Elements According to Toxicity and Availability (after Wood, 1974)

Noncritical			Toxic but very insoluble or very rare		Very toxic and relatively accessible [a]		
Na	C	F	Ti	Ga	**Be**	**As**	Au
K	P	Li	Hf	La	Co	**Se**	**Hg**
Mg	Fe	Rb	Zr	Os	**Ni**	Te	**Tl**
Ca	S	Sr	W	Rh	**Cu**	Pd	**Pb**
H	Cl	(Al)[b]	Nb	Ir	**Zn**	**Ag**	**Sb**
O	Br	Si	Ta	Ru	Sn	**Cd**	Bi
N			Re	Ba	**Cr**	Pt	

a) Priority pollutants for treatment plants (expanded list of 129 compounds or elements considered for categorial standards by EPA Anthony & Breimhurst, 1981) are marked in bold face.

b) Aluminium is toxic for aquatic and terrestrial biota when mobilized at low pH-values.

There are many **pathways or routes** by which aquatic or terrestrial biota and in particular humans are exposed to metallic compounds, and these are changing as society uses more or less of the metal or changes the chemical form of the metal in the environment. To assess which elements may be of concern, four criteria have been proposed: (i) Has the **geochemical cycle** of the element been substantially perturbed by man, and on what scales? (ii) Is the **element mobile** in geochemical processes because of either its volatility or its solubility in natural waters, so that the effect of geochemical perturbations can propagate through the environment? (iii) What is the degree of **public health concern** associated with the element? (iv) What are the critical pathways by which the most **toxic species** of the element can reach the organ in man which is most sensitive to its effect?

Of the elements listed in Table 2-3 global perturbations are most dramatically seen for **lead**. Present-day levels of lead in Americans and Europeans are probably two to three orders of magnitude higher than those of pretechnological humans, as evidenced from studies on blood lead concentrations in remote populations. Changes on a regional scale are typically found for **aluminium** mobilization in soils and waters of low buffer capacity affected by acid precipitation; despite insignificant anthropogenic inputs of aluminium increased solubility will induce toxic effects on both terrestrial and aquatic biota. **Chromium** usually represents examples of only local significance; here, elemental species exhibit characteristic differences, in that the hexavalent form is more toxic than the trivalent form. Other elements, such as lead and **mercury** in Table 2-3, may be mobilized by the biotic or abiotic formation of organometallic compounds. Accumulation of methyl-mercury in seafood, probably the most critical pathway of a metal to humans, has affected several thousand cases of poisoning incidents in Japan (Takeuchi *et al.*, 1959). The first catastrophic event of **cadmium** pollution, affecting the "Itai-Itai" disease to inhabitants in the Jintsu River area of Japan during 20 years after the Second World War, has been caused by effluents from zinc mine wastes, which flooded low-lying rice field areas (Kobayashi, 1971). Cadmium pollution has been recorded from sediment studies in different regions of the world and is related to various sources (Table 2-4); particularly high concentrations have been measured in the Hudson River Estuary, New York (nickel-cadmium battery factory), in the Hitachi area near Tokyo (braun tube factory), in Palestine Lake/Indiana (plating industry), in Sörfjord/Norway and Derwent Estuary/Tasmania (smelter emissions) and from the Neckar River/FRG (pigment factory).

Table 2-3 Perturbation of the Geochemical Cycles of Selected Metals by Society (Examples after Andreae *et al.* in Nriagu, 1984)

Element	Scale of Perturbation			Diagnostic Environments	Mobilizing Mechanisms	Critical Pathway
	Global	Reg.	Local			
Lead	+	+	+	Ice, Sediment	Volatilization	Air, Food
Aluminium	-	+	-	Water, Soil	Solubilization	Water
Chromium	-	-	+	Water, Soil	Solubilization	Water
Mercury	(-)	+	+	Fish, Sediment	Alkylation	Food (Air)
Cadmium	(-)	+	+	Soil, Sediment	Solubilization	Food

In **mineralized zones**, particularly in regions of sulfidic lead-zinc mineralizations, significant accumulations of cadmium takes place in the river sediments. The area around Coeur d'Alène River in Idaho, the Tennessee River near Knoxville, and many rivers in Wales, Southeast England are examples of the lead-zinc mining effect.

Table 2-4 Cadmium in Polluted River Sediments. For References See Förstner & Wittmann (1979)

	Cadmium (ppm)	Source	Reference
North America			
Susquehana River	1.68		Malo (1977)
Harrisburg, PA			
Grand River, MI	3.5	Domestic effluents	Fitchko and Hutchinson (1975)
Grand Calumet River, IN	9.7		Hess and Evans (1972)
	3.1–7.9	Domestic effluents	Romano (1976)
Murderkill River, DE	0.8–8.7		Bopp et al. (1973)
Illinois River	2.0		
	(0.2–12.1)		
Rideau River, Ont.	0.3–15	Mixed effluents	Agemian and Chau (1977)
Lake Cayuga tributaries	15.6		Kubota et al. (1974)
Saginaw River, MI	28		Hess and Evans (1972)
Coeur d'Alene River	Max. 80	Mine effluents	Maxfield et al. (1974)
Milwaukee River, WI	16.6	Industrial effluent	Fitchko and Hutchinson (1975)
	Max. 149		
Tennessee River	Max. 227	Mine effluents	Perhac (1972)
Los Angeles River, CA	860	Sewage effluent	Chen et al. (1974)
Hudson River estuary, NY	2.3	Ni-Cd-battery	Vaccaro et al. (1972)
Foundry Cove, NY	Max. 50,000	factory	Kneip et al. (1974)
South Africa, Australia, Japan			
Gold mine drainage,	0.21	Domestic and mine	Wittmann and Förstner (1976a)
South Africa	(0.05–1.0)	effluents	
Jukskei River, South Africa	0.25–4.9		Wittmann and Förstner (1976b)
Molonglo River, Australia	0.8–3.3	Mining wastes	Australian Government Technical Commission (1974)
Tamar River, Tasmania	3.6	Mine effluents	Ayling (1974)
	(< 0.1–6.0)		
South Esk River, Tasmania	Max. 153	Mine effluents	Tyler and Buckney (1973)
Tama River, Tokyo	0.7–9.8		Suzuki et al. (1975)
Jintsu River, Toyama Pref.	3.27	Mine effluents	Goto (1973)
	(0.16–5.0)		
Takahara River	121	Mine effluents	Kiba et al. (1975)
(near Kamioka mine)	(4.1–238)		
Rivers around Himeji City	0.56–10.4		Azumi and Yoneda (1975)
(W of Osaka)	Max. 129		
Rivers in the Hitachi area, northeast Tokyo	Max. 368	Braun tube factory	Asami (1974)
Israel and Europe			
Gadura River	Max. 123	Battery factory	Kronfeld and Navrot (1975)
(Bay of Haifa, Israel)			
Lake Geneva tributaries,	1.4		Vernet (1976)
Switzerland	(0.09–12.4)	Industrial effluents	Viel et al (1978)
Upper Rhône, Switzerland	0.1–73		Ribordy (1978)
Elbe, FRG	2.9–19.9		Lichtfuss & Brümmer (1977)
Sajo River, Hungary	Max. 20		Literathy and Laszlo (1977)
Blies, Saar, FRG	0.5–24.0	Industrial effluents	Becker (1976)
Bavarian rivers, FRG	< 0.05–29.2	Industrial effluents	Bayerische Landesanstalt für Wasserforschung (1977)
Main River, FRG	17–151		Schleichert and Hellmann (1977)
Ginsheimer Altrhein, FRG	2–95	Industrial effluents	Laskowski et al. (1975)
Neckar River, FRG	Max. 320	Pigment factory	Förstner and Müller (1974)
River Conway, GB	21	Mine effluents	Thornton et al. (1975)
mineralized areas	(3-95)		
River Tawe, GB	Max. 355	Metal processing	Vivian and Massie (1977)
Sava basin, Yugoslavia	Max. 66	Industrial effluents	Štern and Förstner (1976)
Voglajna River			
Stola River, Poland	Max. 116	Mine effluents	Pasternak (1974)
Meuse River, Belgium	Max. 230	Industrial effluents	Bouquiaux (1974)
Vesdre River, Belgium (near Liège)	Max. 430		Bouquiaux (1974)

Metal pollution in surface waters, which caused considerable public concern during the 1970's, seems to have peaked in critical examples, mainly as a result of improvements in **industrial wastewater treatment**. Problems with metals which are dispersed in the environment still exist both on a local and regional scale. Large quantities of **waste materials** on land and in aquatic systems represent long-term reservoirs for the release of metals. Among the various factors enhancing metal mobility acid interactions - both from **acid precipitation** and oxidation of sulfide minerals in mine wastes and dredged sediments - deserved particular attention due to the fact that ionic species predominates, which is readily available for biological uptake. In view of the close linkage of the metal cycles in aquatic systems with the air and soil environments (Stumm, 1986), **indirect effects** from large-scale perturbations can be expected in both surface and groundwaters (Andreae *et al.*, in Nriagu, 1984): Agricultural and residential patterns are changing the rates of continental weathering and erosion. Deforestation on a large scale and draining of major marshes will reduce the number of sinks available for the mobilized metals. The acidification of soils enhances the rate of podzolization and consequently the release of many metals which are otherwise rather immobile; low pH-values must be expected to alter soil microbial ecology, e.g. the prevalence of acidophilic microorganisms, which will in turn influence biogeochemical processes in soils, including related metal cycles (Salomons & Förstner, 1984).

Since the general objective in pollution control is **containment**, processes involving **dispersion** of metals should be avoided (Chapter 5). While emission control of fossil fuel burning, smelting and cement production is of prime importance, liming of soils and waters as well as appropriate techniques for solid waste disposal, such as recultivation of mine spoil heaps could reduce fluxes and biological availability of toxic metals. Agricultural use of metal-rich sewage sludge and fertilizer has to be minimized, as well as injection of wastewater, not only for the sake of food but also for groundwater quality. Future efforts will not only be aimed for chemically stabilizing critical compounds in their deposits but in particular for **recycling** valuable components in waste materials. For example, concentrations such as lead, zinc, and silver in certain fractions of metal-bearing wastes - including metal sludge from electroplating, heat treatment, inorganic pigment manufacture, lime treatment of spent pickle liquor and emission control sludge from waste combustion could well compete with natural resources of some elements (Ball *et al.*, 1987).

2.3. Organic Chemicals

There are a wide range of **organic compounds synthesized** by man for various uses in modern society. The total number synthesized exceeds 100,000 with an estimated 60,000 in common use and approximately 1,000 added per year (Maugh, 1978); it has been estimated, that approximately 1,000 substances are manufactured in quantities that potentially could pollute the globe if released. Among the manifold examples of adverse effects caused by organic compounds actual cases include chlorinated pesticides such as **DDT** which affect egg shell thinning in certain populations of birds (Jensen *et al.*, 1969), **polychlorinated biphenyls** (contamination of rice cooking oil caused over 1000 victims in Japan in 1968 exhibiting symptoms such as nausea, headache, diarrhea and acne, and inducing birth defects; Kurzel & Centrolo, 1981), and solvents like **tetrachloroethylene** in drinking water (Anon., 1980).

Sediment concentration data of major groups of organic contaminants - aliphatic and aromatic hydrocarbons, chlorinated pesticides, phenols, polychlorinated biphenyls and -dibenzodioxins - are summarized in Table 2-5 (the book by Moore & Ramamoorthy contains information on physico-chemical properties, production, uses and discharges, and on toxicity).

Table 2-5 Concentrations of Typical Organic Pollutants in Sediments (Compiled After Moore & Ramamoorthy, 1984)

Aliphatic Hydrocarbons

Because of their volatility most aliphatics - e.g. chloromethane, chloroform, carbon tetrachloride, trichloroethylene, tetrachloroethylene - occur at low or nondetectable levels in sediments.

Aromatic Hydrocarbons - Monocyclics

Benzene, toluene, and some of their derivates are both moderately volatile and soluble in water. Consequently, large-scale sorption to sediments does not occur. Substitution by Cl or N in compounds such as dichlorobenzene, hexachlorobenzene, and dinitrotoluene affects relative high concentrations in sediments near specific industries (see 2.4).

Aromatic Hydrocarbons - Polycyclics

Unsubstituted aromatics such as benzo[a]pyrene, fluoranthene, pyrene, and anthracene are usually found at higher levels associated with heavy industrial activity. Davies *et al.* (1981) showed that very high levels of PAH can be found in sediments around drilling platforms.

---

continued Table 2-5

Chlorinated Pesticides

Highest DDT concentrations averaging 94.000 ng/g were recorded in sediments from the Southern California Bight near Los Angeles, resulting in reproductive failure of local birds, mammals, and fish (Young & Heesen, 1978). Because the half-life of many of these agents is long, sediments will continue to be a source of contamination for many years to come.

Phenols

Eder & Weber (1980) reported that pentachlorophenol residues in the Weser Estuary (FRG) averaged 13 ng/g dry weight. Discharges from a wood preservation plant into the coastal waters of British Columbia resulted in an average pentachlorophenol concentration of 65 ng/g (Jones, 1981); tetra- and trichlorophenol were found with 96 and 26 ng/g, respectively, at the same location. The accidental release of PCP into a small stream produced peak sediment levels of 1300 ng/g (Pierce *et al.*, 1977)

Polychlorinated Biphenyls

Although concentrations of PCBs in some major aquatic systems have declined in recent years, industrial and municipal sources still contribute significantly to the total burden in such areas. Frank *et al.* (1981) reported sediment residues of 10-20 ng/g dry weight in Lake Michigan near Chicago. In the vicinity of waste outfalls, residues may range from 2000 to >500.000 ng/g (Elder *et al.*, 1981).

Polychlorinated Dibenzo-p-Dioxins

Historical fluxes of dioxins and dibenzofurans to sediment cores from Lake Erie and Siskiwit Lake (Isle Royale; Czuczwa & Hites, 1986) suggest that the incineration of chloro-aromatics has been an important source of dioxins and dibenzofurans. In storm sewer and creek sediment samples from the Love Canal chemical dumpsite area in Niagara Falls, N.Y. concentrations of 2,3,7,8-tetrachlorodibenzo-*p*-dioxin from 0.9 to 312 ng/g were found (Smith *et al.*, 1983); this compound in the Love Canal area may be associated with a heavy, chlorinated, oily residue.

In natural waters, the **fate** of hydrophobic organic chemicals, those compounds with a low (less than a few parts per million) solubility, is highly dependent on their "sorption" to suspended particulates. Removal of organic chemicals from solution and their partitioning onto particulates may not be so much of a sorption process but one of exclusion and solubilization of the organic chemical into an organic lipid-like surface layer on particulates. Several mechanisms are involved in particular so-called **"hydrophobic bonding"** where organic molecules are "squeezed" from their coordination with water molecules (Calvet, 1980). As water solubility increases, the **octanol/water partition coefficient** (generally used as a measure of the "lipophility" of a compound) decreases or the compound becomes less soluble in lipid-like organic media (Figure 2-4; after Chiou et al., 1977).

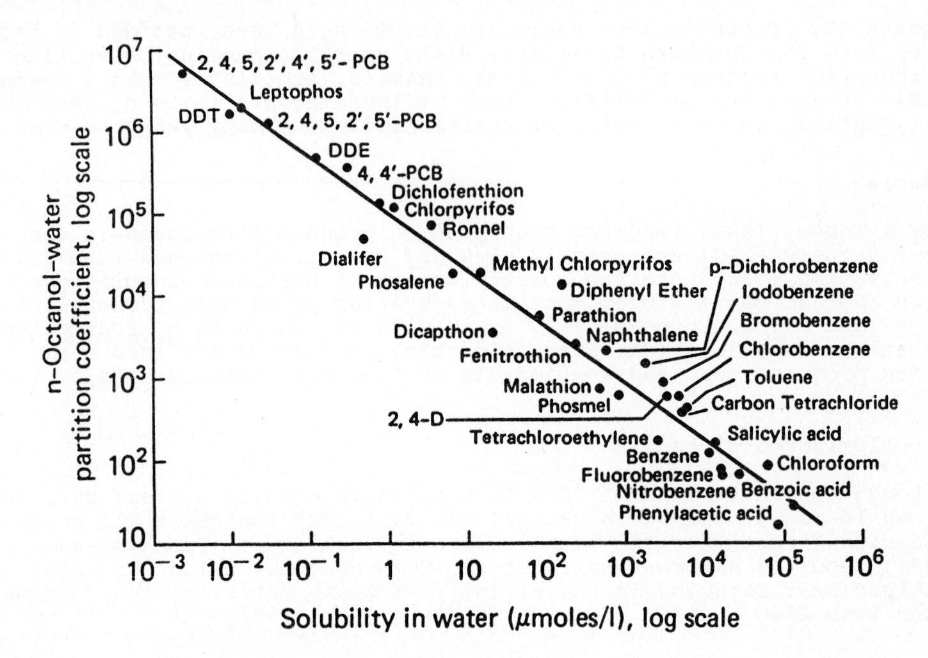

Figure 2-4: Lipophility of Organic Chemicals: Relation between *n*-Octanol/Water Partition Coefficients and Aqueous Solubility (From Chiou *et al.*, 1977)

The **sorption** of organic chemicals on solid surfaces is dependent on their functional groups, the size and shape of the molecule and - if there is any - their charge. Based on these properties, the following **categories** can be set up:

(a) **Cationic or basic compounds,** such as herbicides paraquat and diquat, which interact with negatively charged particles and are strongly or irreversibly bound;

(b) **acidic compounds,** such as the herbicides 2,4 D or 2,4,5 T (phenoxy acids), which were repelled by the negative charge of mineral or organic material;

(c) **non-polar, volatile substances,** such as toluene, which are weakly interacting with particles by hydrophobic bonding; and

(d) **non-polar, non-volatile organic substances,** such as HCH, HCB and DDT, showing stronger hydrophobic bonding forces with particles than (c), increasing with decreasing water solubility.

Attempts have been made to develop methods of estimating degree of sorption for any organic compound on any sorbing substrate. Karickhoff *et al.* (1979) were the first to develop a prediction of **partition of organic chemicals** between water and sediments. Given a specific concentration of the chemical in water and a specific concentration of suspended sediment, a specific partition coefficient (K_D) can be measured. Pavlou & Dexter (1980) summarized partition coefficients for classes of organic pollutants (total number of compound examined: 160) that occur in aquatic ecosystems; it is shown in Table 2-6, that the range from high to low **K-values** follows a shift from non-polar to semi-polar to polar characteristics, which primarily result in changes in water solubility.

Table 2-6 Estimated Range of Sediment/Water Partition Coefficients for Various Pesticide Classes (after Pavlou & Dexter, 1980)

Pesticide Type	K-Value	Characteristics of Adsorption
Organochlorines		Few polar moieties; hydrophobic and Van
Aromatic	10^5-10^3	der Wasls interaction; induction effects
Aliphatic		substituents, non-conjug. double-bonds)
Organophosphates		
Aliphatic deriv.	$5x10^2$-10^1	Active polar moieties (electron-rich
Phenyl derivates	10^3-10^2	heteroatoms, atoms, acidic hydrogens,
Heterocyclic	$5x10^2$-50	heterocyclic nitrogen)
Carbamates		Highly polar; enhanced solubility;
Methyl carbamates	$5x10^2$- 2	reduced adsorption compared with
Thiocarbamates	$5x10^2$-50	organochlorines
Nitroanilines	$1x10^3$-50	Nonconjugated polar groups; large molecules; strong hydrophobic forces
Triazines	8 - 1	Solar ionizable groups; hydrogen bonding with water; low adsorption

For **non-polar organic contaminants** (category "d") interactions can be described empirically by simple relationship to the content of organic matter (Karickhoff, 1981). The partition coefficient normalized to organic carbon contents of the solids (K_{OC} = K_D/fraction of organic carbon) should be highly invariant over a wide range of substrate types. These findings can be used for establishing "sediment quality criteria" for neutral organic compounds on the basis of partition coefficients to the respective water quality standards (see Chapter 6).

2.4. Case Study: Pollutants in Sediments of the Niagara River

The Niagara River flows from Lake Erie to Lake Ontario and forms part of the boundary between Canada and the United States. Niagara Falls lies 35 km upstream from the in-flow site of the Niagara River to Lake Ontario. The Niagara River is one of the most polluted waterways in the world in terms of the quantity and diversity of chemicals transported to Lake Ontario (Allan *et al.*, 1983).

Input of contaminants from Niagara River into Lake Ontario were in the headlines already at the beginning of the seventies in the context of the **mercury problems** in the Great Lakes Region, when commercial fishing was banned in the waters of the St.Clair River-Lake Erie System. Consequently, the classical survey on mercury in surficial sediments of Lake Ontario by Thomas (1972) was conducted, based on the analysis of several hundred samples. Figure 2-5 shows a distinct grouping of elevated mercury levels near the southern bank of the lake, especially close to the mouth of the Niagara River. Further investigations carried out by Fitchko & Hutchinson (1975) have in fact shown that the dispersion pathways point to the Niagara River as the prime source of mercury input to Lake Ontario. The eastward extension of the Niagara mercury "plume" is probably due to coastal current, whereas the enrichment of mercury at the mouth of the Genesee and Oswego Rivers appears to result from the contamination within their catchment areas.

In order to understand the full extent of the mercury problem in these times, one has only to consider the enormous **loss rates**. From the total of 2865 tons of mercury purchased in the U.S. in 1968, 76% or 2160 tons were lost to the environment. According to calculations of Kemp *et al.* (1974), the Lake Ontario reservoir contained a mass of 500 to 600 metric tons of "excess" mercury, i.e. discharged from anthropogenic sources. With the improvements in the methods of chlor-alkali electrolysis and by subsequent purification of waste streams the mercury loss has been reduced from 100 g per metric ton of manufactured chlorine to approx. 2 g per ton or less (Anon., 1973). The effect of these measures can be seen from concentration profiles of mercury in **sediment cores** taken off the mouth of Niagara River by Mudroch (1983), where a very distinct decrease from formerly approx. 4-7 μg Hg/g to less than 1 μg Hg/g in recent years has occurred (Figure 2-6).

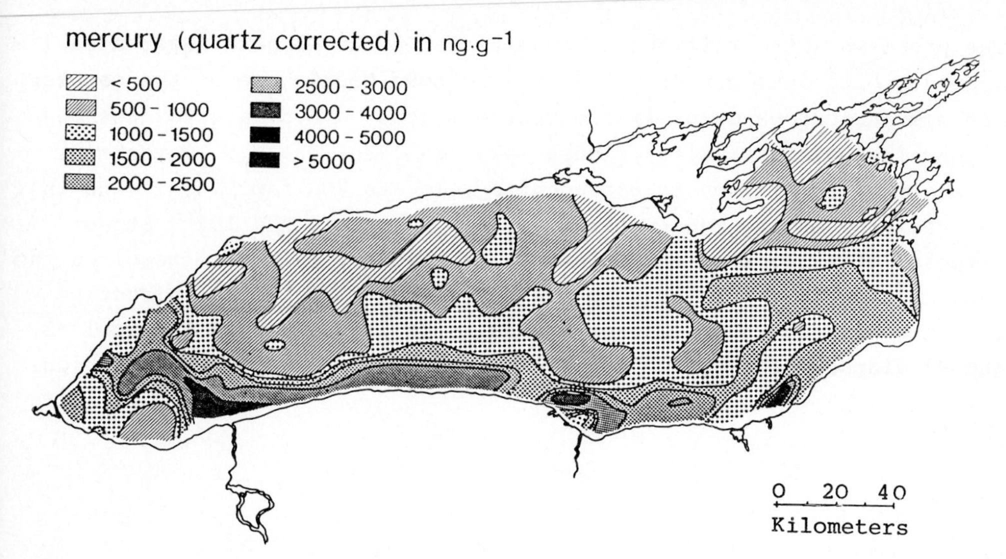

Figure 2-5: The Distribution of Quartz-Corrected Mercury in Lake Ontario, Uppermost 3 cm of Sediment (Thomas, 1972)

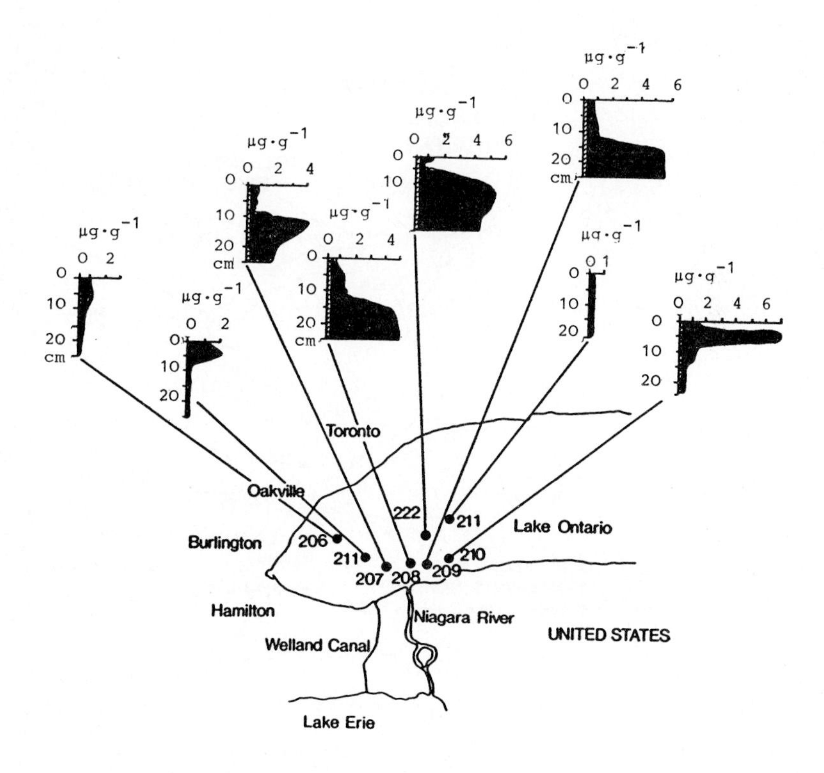

Figure 2-6: Concentration Profiles of Mercury in Sediment Cores from the Western Basin of Lake Ontario (Mudroch, 1983)

The problems with toxic organic chemicals in the Niagara River area roused public interest with the case of **Love Canal**, one of the largest and most notorious dumps in the United States, a 6.5-ha site just upstream from Niagara Falls (Figure 2-7). In 1980, the U.S. Government allocated $ 80 million to permanantly relocate 700 families living in the vicinity of the Canal (Allan, 1986). Nearby is the **102nd Street disposal site**, located in Niagara Falls, U.S.A.; this site required the construction of a bulk-head to prevent wastes being washed directly into the Niagara River. Closer to the Falls is the **Hooker Chemical "S and H" disposal site** which contains some 75.000 tons of waste liquids.

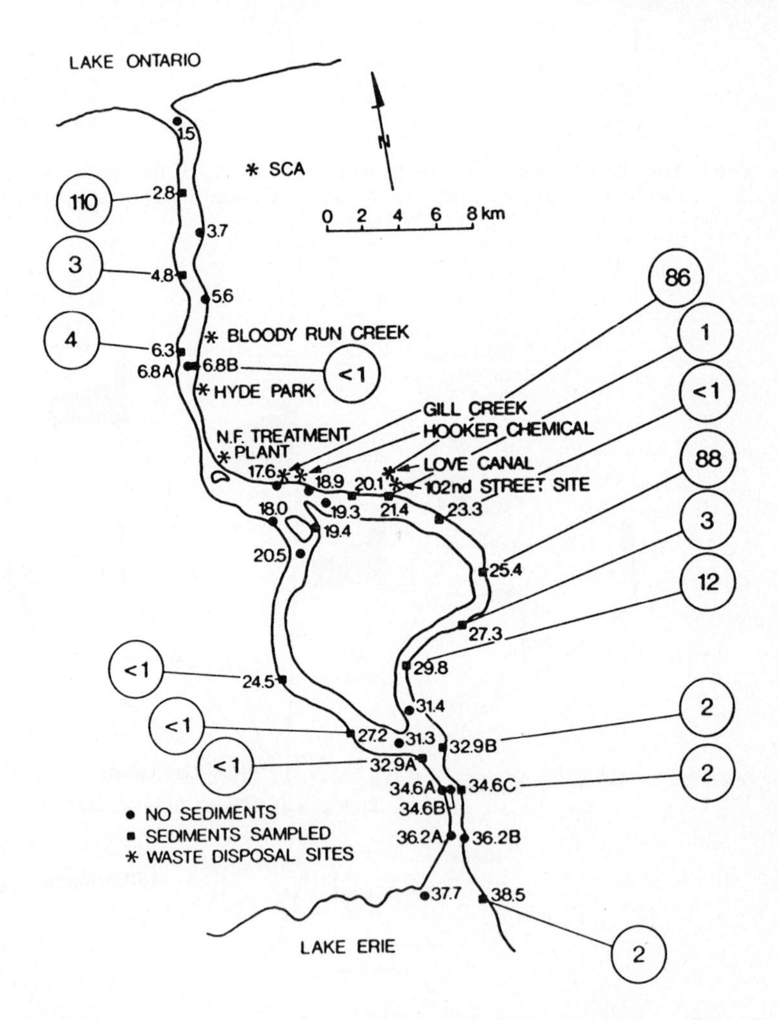

Figure 2-7: Location Map of Waste Disposal Sites and Wastewater Outfalls, and Distribution of Hexachlorobenzene (ng/g) in Bottom Sediments of the Niagara River, May 1981 (Kuntz, 1984)

Below the Falls, the 6.1-ha **Hyde Park disposal site**, adjacent to the Niagara Gorge, was used between 1953 and 1974 to dispose of some 80.000 tons of chemicals. Drainage from the Hyde Park dump followed **Bloody Run Creek** to the Niagara River. The two largest wastewater point sources that have received the most public attention are the outfalls of the **Niagara Falls, U.S.A., Wastewater Treatment Plant** immediately below Niagara Falls and the **SCA Chemical Waste Services Facility** pipeline near Lake Ontario.

In 1977 there were 105 **chemical and allied products** establishment in the Niagara River area. Some of the largest U.S. chemical corporations have plants in the "Niagara Frontier" (Allan *et al.*, 1983); ranked by chemical sales, thirteen of the top 50 are represented in the Niagara area, e.g., Dupont, Union Carbide, Allied, Ashland, and FMC. Among these, the **Hooker Electrochemical Co.** plant received most public attention due to its involvement in the Love Canal affair (Brown, 1981). Hooker Co. was the first in the U.S.A. to produce **chlorobenzenes**, of which the lower chlorinated CBs are widely used in industry as solvents, intermediate compounds and as primary products in pesticides, dyestoffs, plastics, odour control chemicals, etc., whereas the more highly chlorinated CBs, tetra, penta and hexa, have very limited industrial or commercial applications and are mainly produced as unwanted products in the manufacture of the lower chlorinated isomers (Durham & Oliver, 1983). Operations at Niagara Falls began in 1915 with a capacity of 8,200 metric tons/y. Production dramatically increased in the early 1940s, but manufacturers stopped used CB compounds in the production of phenols in the late 1960s and this resulted in a decrease in production after 1970.

The **CB profiles** in sediment cores from Lake Ontario taken by Durham & Oliver (1983) a few kilometers off the inflow of the Niagara River clearly demonstrate this development. In the example of a ^{210}Pb-dated core from this area (August 19, 1981; 70 m water depth) given in Figure 2-8 the historical development of the more soluble, lower chlorinated chlorobenzene compounds **1,3-DCB** and **1,4-DCB** (see Table 2-7) exhibits two significant peak concentrations during the Second World War (350 µg/g and 850 µg/g dry matter, respectively) and in the 1960's (1000 µg/g for both substances); since 1968, concentrations have gradually decreased to less than 200 µg/g (Durham & Oliver, 1983). This coincides with the suspended sediment concentrations of 95 µg/g 1,3-DCB and 180 µg/g 1,4-DCB, respectively, in 28 samples taken from Niagara River at Niagara-on-the-Lake in 1980 (Kuntz & Warry, 1983).

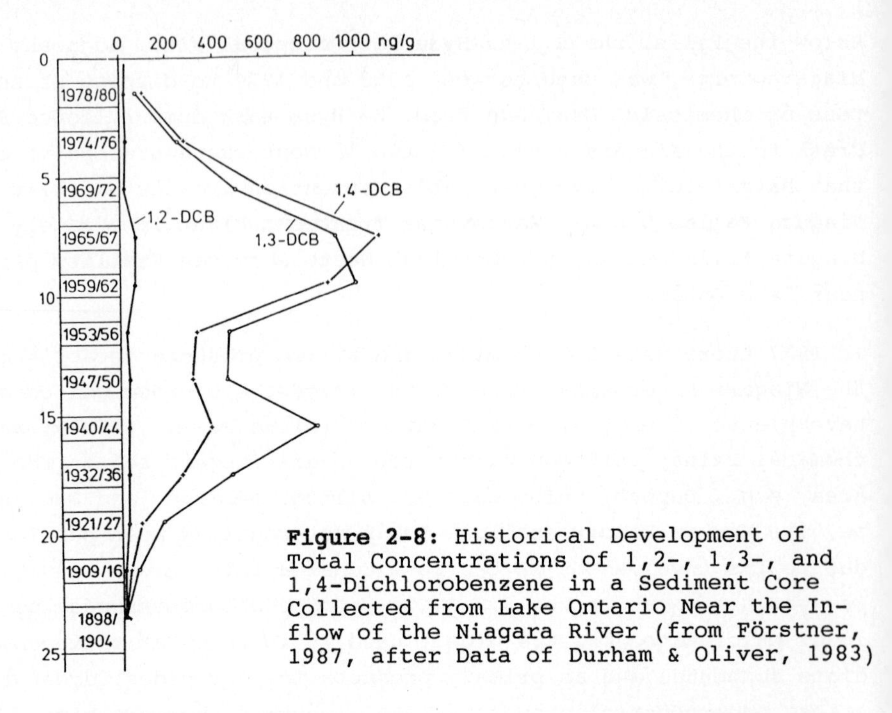

Figure 2-8: Historical Development of Total Concentrations of 1,2-, 1,3-, and 1,4-Dichlorobenzene in a Sediment Core Collected from Lake Ontario Near the Inflow of the Niagara River (from Förstner, 1987, after Data of Durham & Oliver, 1983)

Table 2-7 Percent Particulate-Transported Organic Chemicals to Lake Ontario by the Niagara River (After Allan, 1986). [a] 1979-1981 (Kuntz *et al.*, 1982); [b] 1982 (Oliver & Nicol, 1982)

Chemical	Total load (kg/yr)	Total in suspended sediments (kg/yr)	**Percent of total in suspended sediment**
Total PCBs [a]	2200	850	**39 %**
Total DDT	20	41	**100 %**
Mirex	20	8	**40 %**
Lindane	400	2	**<1 %**
1,2-Di-CB [b]	1000	2	**<1 %**
1,4-Di	1000	2	**<1 %**
1,2,3-Tri	200	2	**1 %**
1,2,4-Tri	200	27	**13 %**
1,2,3,4-Tetra	200	27	**13 %**

The **quantity** of toxic organic chemicals transported by the Niagara River at the beginning of the 80's is estimated for typical examples in Table 2-7. PCB's, of which approximately one hundred times more - compared to DDT and Mirex - are discharged through the Niagara River system, are distinctly associated with particulate matter; relative high discharges of Lindane, on the other hand, are predominantly found with the aqueous phase. Regarding the chlorobenzenes in Table 2-7, lower chlorinated compounds are mainly transported in the water phase, whereas for the trichlorobenzenes typical differences can be observed with 1,2,3-Tri- and 1,2,4-Tri-isomers, of which only the latter molecule exhibits a distinct tendency for particle-associations (Chapter 4.4).

Typical enrichment of organic chemicals in the **aquatic food chain** of Lake Ontario has been observed for **DDT** and **PCBs**, which exhibit concentrations up to 8 μg/g and 17 μg/g, respectively, in fish; on a lower concentration level - mainly due to the higher solubility of these compounds - similar effects can be seen for **mirex** and **lindane** (Table 2-8). Chlorobenzenes, in relation to sediment data, do not exhibit food chain enrichment, except for **hexachlorobenzene**, which is significantly concentrated in planktonic and benthic organisms. However, there is a distinct accumulation of all chlorobenzene isomers by organisms from the water phase, and it has been stressed by Oliver & Nicol (1982) as a typical feature that Lake Ontario fish contains much higher concentrations of CB's than fish from the other Great Lakes.

Table 2-8 Range of Concentrations (in ng/g) of Organic Contaminants in Sediments and Biota of Lake Ontario (Allan, 1986, after Oliver & Nicol, 1982, Fox *et al.*, 1983, and Strachan & Edwards, 1984)

Chemical	Suspended Sediment	Bottom Sediment	Plankton	Benthos	Fishes
PCBs	600-6100	110-1600	110-6100	470-9000	1378-17000
Total DDT	40	25- 218	63- 72	440-1088	620- 7700
Mirex	15	144	ND- 12	41- 228	50- 340
Lindane	1- 12	46	12	NA	2- 360
1,4-DCB	100- 600[a]	33-1300	71- 370	170- 630	4[b]
1,2,4-TCB	33- 210[a]	59- 220	6- 330	20- 81	5[b]
1,2,4,5-TeCB	14- 160[a]	36- 210	8- 210	21- 69	5[b]
QCB	17- 400[a]	26- 120	14- 210	21- 81	16[b]
HCB	17- 460[a]	62- 840	90-1400	63-1200	127[b]

ND = not detected; NA = not analyzed

(a) Size fractionated suspended solids (3 samples, 6 fractions) from Niagara River at Niagara-on-the-Lake (Fox *et al.*, 1983)

(b) Lake Ontario lake trout (age 6+ years, lipid content 23%), caught off Niagara River (Oliver & Nicol, 1982)

References to Section 2.1 "Historical Development"

Alderton, D.H.M. (1985) Sediments. In *Historical Monitoring,* MARC Technical Report 31, pp. 1-95. London: Monitoring and Assessment Research Centre/University of London.

Allan, R.J. (1971) Lake sediment, a medium for regional geochemical exploration of the Canadian Shield. *Can. Inst. Min. Met. Bull* 64, 43-59.

Allan, R.J. (1974) Metal contents of lake sediment cores from established mining areas: An interface of exploration and environmental geochemistry. *Geol. Surv. Can.* 74-1/B, 43-49.

Chapra, St.C. (1977) Total phosphorous model for the Great Lakes. *J. Env. Eng. Div. Am. Soc. Civil Eng.* 103, 147-161.

Dahlberg, E.C. (1968) Application of selective simulation and sampling technique to the interpretation of stream sediment copper anomalies near South Mountain, Pa. *Econ. Geol.* 63, 409-417.

Hakanson, L. (1977) *Sediments as Indicators of Contamination -Investigations in the Four Largest Swedish Lakes.* SNV PM 839, NLU Report 92, 159 p. Uppsala: Naturvardsverkets Limnologiska Undersökning.

Hawkes, H.E. & Webb, J.S. (1962) *Geochemistry in Mineral Exploration.* 415 p. New York: Harper & Row.

Holdrinet, M. *et al.* (1978) Mirex in the sediments of Lake Ontario. *J. Great Lakes Res.* 4, 69-74.

Huggett, R.J., Nichols, M.M. & Bender, M.E. (1980) Kepone contamination of the James River estuary. In *Contaminants and Sediments,* ed. R.A. Baker, Vol. I, pp. 33-52. Ann Arbor: Ann Arbor Sci. Publ

Meiggs, T.O. (1980) The use of sediment analysis in forensic investigations and procedural requirements for such studies. In *ibid.* pp. 297-308.

Müller, G. (1981) Heavy metals and other pollutants in the Environment. A chronology based on the analysis of dated sediments. In Proc. Int. Conf. *Heavy Metals in the Environment, Amsterdam,* ed. W.H.O. Ernst, pp. 12-17. Edinburgh: CEP Consultants.

Santschi, P.H. *et al.* (1988) Chernobyl radionuclides in the environment: Tracers for the tight coupling of atmospheric, terrestrial, and aquatic geochemical processes. *Environ. Sci. Technol.* 22, 510-516.

Stumm, W. & Baccini, P. (1978) Man-made chemical perturbation of lakes. In *Lakes - Chemistry, Geology, Physics,* ed. A. Lerman, pp. 91-126. New York: Springer-Verlag.

Turk, J.T. (1980) Applications of Hudson River basin PCB-transport studies. In *Contaminants and Sediments,* ed. R.A. Baker, pp. 171-184. Ann Arbor: Ann Arbor Sci. Publ.

Williams, J.D.H., Murphy, T.P. & Mayer, T. (1976) Rates of accumulation of phosphorous forms in Lake Erie sediments. *J. Fish Res. Board Can.* 33, 430-439.

References to Section 2.2 "Metals"

Anthony, R.M. & Breimhurst, L.H. (1981) Determining maximum influent concentrations of priority pollutants for treatment plants. *J. Water Pollut. Control Fed.* 53, 1457-1468.

Ball, R.O. *et al.* (1987) Economic feasibility of a state-wide hydrometallurgical recovery facility. In *Metals Speciation, Separation and Recovery*, eds. J.W. Patterson & R. Passino, pp. 690-709. Chelsea/MI: Lewis Publ.

Förstner, U. & Müller, G. (1973) Heavy metal accumulation in river sediments: A response to environmental pollution. *Geoforum* 14, 53-61.

Förstner, U. & Wittmann, G.T.W. (1979) *Metal Pollution in the Aquatic Environment.* 486 p. Berlin: Springer-Verlag.

Kobayashi, J. (1971) Relation between the "Itai-Itai" disease and the pollution of river water by cadmium from a mine. Proc. 5th Int. Conf. on *Advanced Water Pollution Research*, San Francisco and Hawaii, Vol. I-25, pp. 1-7. Oxford: Pergamon Press.

Moore, J.W. & Ramamoorthy, S. (1984) *Heavy Metals in Natural Waters - Applied Monitoring and Impact Assessment.* 268 p. New York: Springer-Verlag.

Nriagu, J.O. (Ed.)(1984) *Changing Metal Cycles and Human Health.* Dahlem-Konferenzen, Life Sci. Res. Rep. 28, 367 p. Berlin: Springer.

Salomons, W. & Förstner, U. (1984) *Metals in the Hydrocycle.* 349 p. Berlin: Springer-Verlag.

Stumm, W. (1986) Water, an endangered ecosystem. *Ambio* 15, 201-207.

Takeuchi, T. *et al.* (1959) Pathological observations of the Minamata disease. *Acta Pathol. Jpn. Suppl.* 9, 769-783.

Wood, J.M. (1974) Biological cycles for toxic elements in the environment. *Science* 183, 1049-1052.

Wood, J.M. & Wang, H.-K. (1983) Microbial resistance to heavy metals. *Environ. Sci. Technol.* 17, 582A-590A.

References to Section 2.3 "Organic Chemicals"

Anonymus (1978) *Cleaning Our Environment - A Chemical Perspective.* 457 p. Washington D.C.: American Chemical Society.

Anonymus (1980) Groundwater strategies. *Environ. Sci. Technol.* 14, 1030-1035.

Calvet, R. (1980) Adsorption-desorption phenomena. In *Interactions between Herbicides and the Soil*, ed. R.J. Hance, pp. 1-29. London: Academic Press.

Chiou, C.T. et al. (1977) Partition coefficient and bioaccumulation of selected organic chemicals. *Environ. Sci. Technol.* 11, 475-478.

Czuczwa, J.M. & Hites, R.A. (1986) Airborne dioxins and dibenzofurans: Sources and fates. *Environ. Sci. Technol.* 20, 195-200.

Davies, J.M., Hardy, R. & McIntyre, A.D. (1981) Environmental effects of North Sea oil operations. *Mar. Pollut. Bull.* 12, 412-416.

Eder, G. and Weber, K. (1980) Chlorinated phenols in sediments and suspended matter of the Weser estuary. *Chemosphere* 9, 111-118.

Elder, V.A., Proctor, B.L. and Hites, R.A. (1981) Organic compounds found near dump sites in Niagara Falls, New York. *Environ. Sci. Technol.* 15; 1237-1243.

Frank, R. *et al.* (1981) Organochlorine insecticides and PCB in surficial sediments of Lake Michigan (1975). *J. Great Lakes Res.* 7, 42-50.

Jensen, S. et al. (1969) DDT and PCB in marine animals from Swedish waters. *Nature* 224, 247-250.

Jones, P.A. (1981) *Chlorophenols and Their Impurities in the Canadian Environment*. Cat. No. En46-3/81-2, 434 p. Ottawa: Department of the Environment, Canada.

Karickhoff, S.W. (1981) Semi-empirical estimation of sorption of hydrophobic pollutants on natural sediments and soils. *Chemosphere* 10, 833-846.

Karickhoff, S.W., Brown, D.S. & Scoll, T.A. (1979) Sorption of hydrophobic pollutants on natural sediments. *Water Res.* 13, 241-248.

Kurzel, R.B. & Centrulo, C.L. (1981) The effect of environmental pollutants on human reproduction, including birth defects. *Environ. Sci. Technol.* 15, 626-640.

Maugh, T.H. (1978) Chemicals: How many are there? *Science* 199, 162.

Moore, J.W. & Ramamoorthy, S. (1984) *Organic Chemicals in Natural Waters - Applied Monitoring and Impact Assessment.* 289 p. New York: Springer-Verlag.

Pavlou, S.P. & Dexter, R.N. (1980) Thermodynamic aspects of equilibrium sorption of persistent organic molecules at the sediment-seawater interface: a framework for predicting distribution in the aquatic environment. In *Contaminants and Sediments*, ed. R.A. Baker, pp. 323-329. Ann Arbor: Ann Arbor Sci. Publ.

Pierce, R.H. jr. *et al.* (1977) Pentachlorophenol distribution in a freshwater ecosystem. *Bull. Envir. Contam. Toxicol.* 18, 251-258.

Smith, R.M. *et al.* (1983) 2,3,7,8-tetrachlorodibenzo-*p*-dioxin in sediment samples from Love Canal storm sewers and creeks. *Environ. Sci. Technol.* 17, 6-10.

Young, D.R. & Heesen, T.C. (1978) DDT, PCB, and chlorinated benzenes in the marine ecosystem off Southern California. In *Water Chlorination - Environmental Impact and Health Effects*, ed. R. Jolley. Vol. 2, pp. 267-290. El Segundo: S.Calif. Coast. Water Res. Proj.

References to Section 2.4 "Case Study: Niagara River"

Allan, R.J. (1986) *The Role of Particulate Matter in the Fate of Contaminants in Aquatic Ecosystems.* National Water Research Institute, Scientific Ser. No. 142, 128 p. Burlington/Ontario: Canada Centre for Inland Waters.

Allan, R.J., Mudroch, A. & Munawar, M. (Eds.) (1983) *The Niagara River - Lake Ontario Pollution Problem. J. Great Lakes Res.* 9, 109-340.

Anonymus (1973) Removal of mercury from chlorine plant effluents. *Environ. Sci. Technol.* 7, 185.

Brown, M. (1981) *Laying Waste - The Poisoning of America by Toxic Chemicals.* 363 p. New York: Washington Square Press.

Durham, R.W. & Oliver, B.G. (1983) History of Lake Ontario contamination from the Niagara River by sediment radiodating and chlorinated hydrocarbon analysis. In: *The Niagara River - Lake Ontario Pollution Problem.* ed. R.J. Allan et al. *J. Great Lakes Res.* 9, 160-168.

Fitchko, J. & Hutchinson, T.C. (1975) A comparitive study of heavy metal concentrations in river mouth sediments around the Great Lakes. *J. Great Lakes Res.* 1, 46-78.

Förstner, U. (1987) Sediment-associated contaminants - an overview of scientific bases for developing remedial options. In *Ecological Effects of In Situ Sediment Contaminants*, eds. R.L. Thomas *et al.* *Hydrobiologia* 149, 221-246.

Fox, M.E., Carey, J.H. & Oliver, B.G. (1983) Compartmental distribution of organochlorine contaminants in the Niagara River and the western basin of Lake Ontario. *The Niagara River - Lake Ontario Pollution Problem.* ed. R.J. Allan et al. *J. Great Lakes Res.* 9, 287-294.

Kemp, A.L.W. *et al.* (1974) Sedimentation rates and recent sediment history of Lakes Ontario, Erie and Huron. *J. Sediment. Petrol.* 44, 469-490.

Kuntz, K.W. (1984) *Toxic Contaminants in the Niagara River, 1975-1982.* Techn. Bull. Water Quality Branch, Ontario Region No. 134, 47 p. Burlington: Canada Centre for Inland Waters.

Kuntz, K.W. & Warry, N.D. (1983) Chlorinated organic contaminants in water and suspended sediments of the lower Niagara River. *The Niagara River - Lake Ontario Pollution Problem.* ed. R.J. Allan et al. *J. Great Lakes Res.* 9, 241-248.

Kuntz, K.W. et al. (1982) *Water Quality Sampling Methods at Niagara-on-the-Lake*. Rept. Water Quality Branch, Ontario Region. Burlington: Canada Centre for Inland Waters.

Mudroch, A. (1983) Distribution of major elements and metals in sediment cores from the western basin of Lake Ontario. *The Niagara River - Lake Ontario Pollution Problem.* ed. R.J. Allan et al. *J. Great Lakes Res.* 9, 125-133.

Oliver, B.G. & Nicol, K.D. (1982) Chlorobenzenes in sediments, water and selected fish from Lakes Superior, Huron, Erie and Ontario. *Environ. Sci. Technol.* 16, 532-536.

Strachan, W.M.J. & Edwards, G.J. (1984) Organic pollutants in Lake Ontario. In *Toxic Contaminants in the Great Lakes,* eds. J.O. Nriagu & M.S. Simmons, pp. 239-264. New York, Wiley.

Thomas, R.L. (1972) The distribution of mercury in the sediments of Lake Ontario. *Can. J. Earth Sci.* 9, 636-651.

Additional Themes (Examples)

"Sediment Contamination from **Mining** and Smelting Activities" (Wales, Southwest England, Sörfjord/Norway, Greenland, Saxony/GDR, Harz/FRG, Poland, Serbia & Kosovo/Yugoslavia, Sudbury/Ontario, British Columbia, Puget Sound/Washington, Missouri, Tennessee, Peru, South Africa, Japan, Philippines, Papua New Guinea, Northern Territory/Australia, Tasmania)

"**Nutrients** (Phosphorous, Nitrogen, Carbon-BOD) in Lacustrine Sediments" (Swiss Lakes; Swedish Lakes; Lakes in Wisconsin, Illinois and Michigan)

"**Estuarine Cycling** of Sediment-Associated Pollutants" (St. Lawrence; U.S. East Coast; Gironde, Scheldt, Rhine, Ems, Weser, Elbe Estuaries)

Case Studies on Lakes and Coastal Areas (Examples)

- Adirondack Mountains & New England Lakes (acidity)
- Wisconsin Lakes (Cu-algicides, As-herbicides)
- Lake St. Clair/Lake Erie (Hg, Cd, nutrients, DDT, PAH)
- Lake Constance (Cd, Pb, Zn, PAH, coprostanol)

- Southern California Bight (DDT, PCBs, nutrients, metals)
- New Bedford Harbour (Cu, Cd, PAH, PCBs, DDE)
- New York Bight/Long Island Sound (nutrients, microbes; PCBs, Cd)
- Chesapeake Bay (metals, DEHP, PAH, paraffin, kepone)
- Baltic Sea/North Sea (DDT, PCB, DEHP, Pb, Zn, Hg, Cd)
- Irish Sea (metals, artificial radionuclides)
- Mediterranean Sea (nutrients, metals, DDT)
- Tokyo and Osaka Bays (metals, PAH, fuel oil, coprostanol)

3. ASSESSMENT METHODS

3.1. Objectives

At the present time, **standard analytic techniques** are available to meet most sampling and analytical requirements for water studies, but the same is not true sediment studies. Sediment sampling and analysis require the use of different techniques and equipment. The areal distribution of the samples is not the same and they are generally taken less frequently in time than for water samples. Nonetheless, sediment analyses are often based upon modified procedures of **water quality** analyses (e.g., Anon., 1978), or upon methods developed either in **soil science** (e.g., Bear, 1964) or for purposes of **mineral exploration** (e.g., Levinson, 1974).

A **long-term program** of sediment studies will normally consist of a series of objectives of increasing complexity, each drawing part of its information from the preceding data base; a typical sequence of objectives may be illustrated as follows, though not all may be required to complete a program (Golterman *et al.*, 1983, pp. 23-26):

a) **Preliminary site characterization** - Low density sampling with limited analytical requirements, to provide a general characterization of an area for which little or no previous information exists.

b) **Identify anomalies** - More detailed sampling and analyses, designed to establish the presence and extent of anomalies.

c) **Establish references** - To create reference points, in the form of some measured parameters, for future comparison.

d) **Identify time changes** - To show trends and variations of sediment data over time, be use of sediment cores or other repeated sediment samplings.

e) **Calculate mass balances** - To account for the addition and subtraction of sediment-related components with an aquatic environment (a complex study), by means of accurate and representative sampling and analysis.

f) **Process studies** - Specialized sampling to improve state of knowledge about aquatic systems, e.g. by supplementary laboratory experiments.

The principal relationships between **sampling objectives** and **type of activities** for water-related studies are summarized in Table 3-1; the third column contains the specific objectives of sediment studies listed above. Sediment **time-series data** (category "d") belong to two different types of activities: Suspended sediment time-series data would be considered as a **monitor surveillance** activity; sediment core samples represent a highly condensed time-series covering relatively long periods between potential samplings, and it may be appropriate to include this sampling objective under **survey** type activities.

Table 3-1 Principal Relationships between Sampling Objectives and Type of Activities in Water- and Sediment-Related Studies (Modified after Golterman et al., 1983)

Type of Activity (UNESCO-WHO, 1978)	GEMS Water Objectives (WHO-Geneva, 1978)	Sediment Objectives (Categories in Text)
Monitor		
Continous standard measurement and ob- observation	Cultural impact on water quality, suitability of water quality for future use	Establish reference point(s) - Cat. (c)
Surveillance		
Continous specific observation and measurement relative to control & management	Observe sources and pathways of specified hazardous substances	Trace sources (spatial)
Survey		
Series of finite duration; intensive, detailed programs for spec. purposes	Determine quality of natural waters	Identify anomalies (category "b") Calculate mass balances (category "e") Study processes (f)

Program objectives largely control the type, density, and frequency of sediment sampling and associated analyses; whereas the **type of environment** (rivers, lakes, estuaries, etc.) largely controls the locations and logistics of sampling. **Logistic factors** include (Golterman *et al.*, 1983, p. 67):

- local availability of sampling platform or vessel
- time available
- access to sampling region
- suitability of survey system to locate sample position
- availability of trained personel and support staff
- availability of equipment
- storage and security
- transport systems
- follow-up capability

The accelerating interest of **environmental agencies** in adopting sediment analysis as an integral component of their programs, substantiates the fact that this type of research has achieved considerable success. It is now possible to outline several useful objectives for sediment sampling programs (Golterman *et al.*, 1983, pp. 67-73) and to present comments about some of their limitations. The **limitations** are generally a function of incomplete knowledge or technique, and some of the major aspects will be treated below.

3.2. Sampling

In general, the density of sampling should be as great as possible, bearing in mind time requirement, capacity of laboratory services, data storage, and handling and processing capabilities. For preliminary or exploratory surveys, the location of survey lines and the position of sample locations depends very much on the size and shape of the water body (Figure 3-1). For more complex surveys, there are numerous types of **sampling patterns** from which to choose, e.g., spot samples, random grids, square grids (including nested and rotated grids), parallel line grids and transverse line grids (with equal or non-equal sampling), and ray grids or concentric arc sampling, each of which offers some particular advantage (Golterman *et al.*, 1983, pp. 76-79).

The suitability of **corers and bottom samplers** have been tested during equipment trials by Sly (1969); Figure 3-2 lists a selection of devices, which have successfully been applied in diffent aquatic environments. For source reconnaissance analysis, fine- to medium-grained bottom deposits from a depth of 15-20 cm can be collected, for example, with an Ekman grab sampler. Material of the upper, flakey, light brown, oxidized layer is generally dissimilar to the layers below it. It is suggested that the chiefly dark layers directly underneath (ca. 1-3 cm depth) are more representative of the pollution situation over the last few years, especially in river deposits exhibiting rapidly fluctuating-

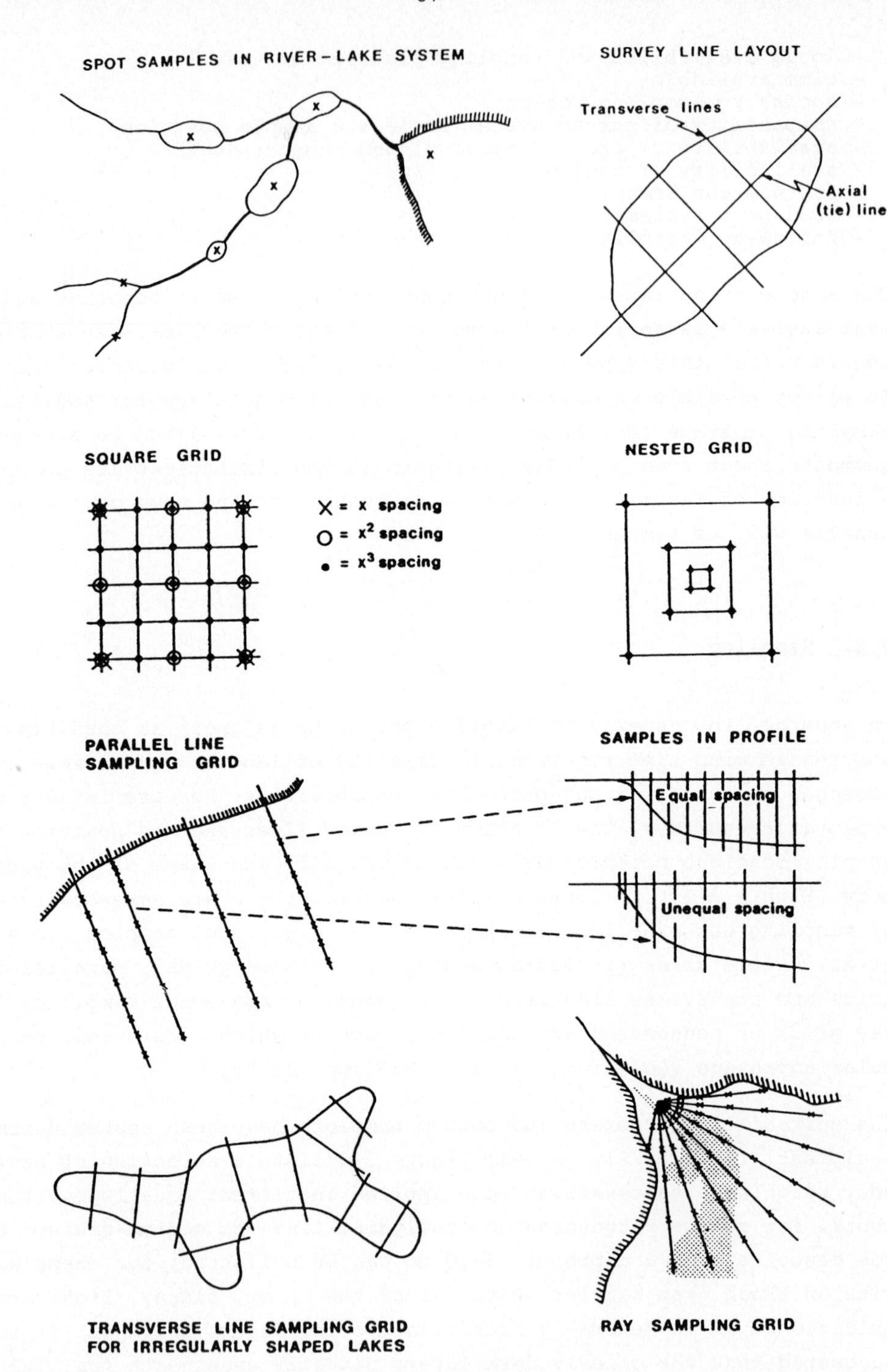

Figure 3-1 Examples of Different Types of Sampling Grid Design (Golterman et al., 1983, p. 79)

sedimentation rates, and should be given priority for subsequent investigations. To complement this, surface sediment (current contamination) as well as a sample from deeper sections (10-20 cm depth) could be examined. In environments with a relatively uniform sedimentation, for example in lakes and in marine coastal basins, where the deposits are fine-grained and occur at a rate of 1 to 5 mm/yr, a more favourable procedure involves the taking of vertical profiles with a gravity or valve corer (see Figure 3-2). A core profile of approximately 1 m covers a historical period of at least 200 years, and its development can be traced by virtue of the pollutant content in the individual layers (see examples in Chapter 2, sections 1 and 4, respectively).

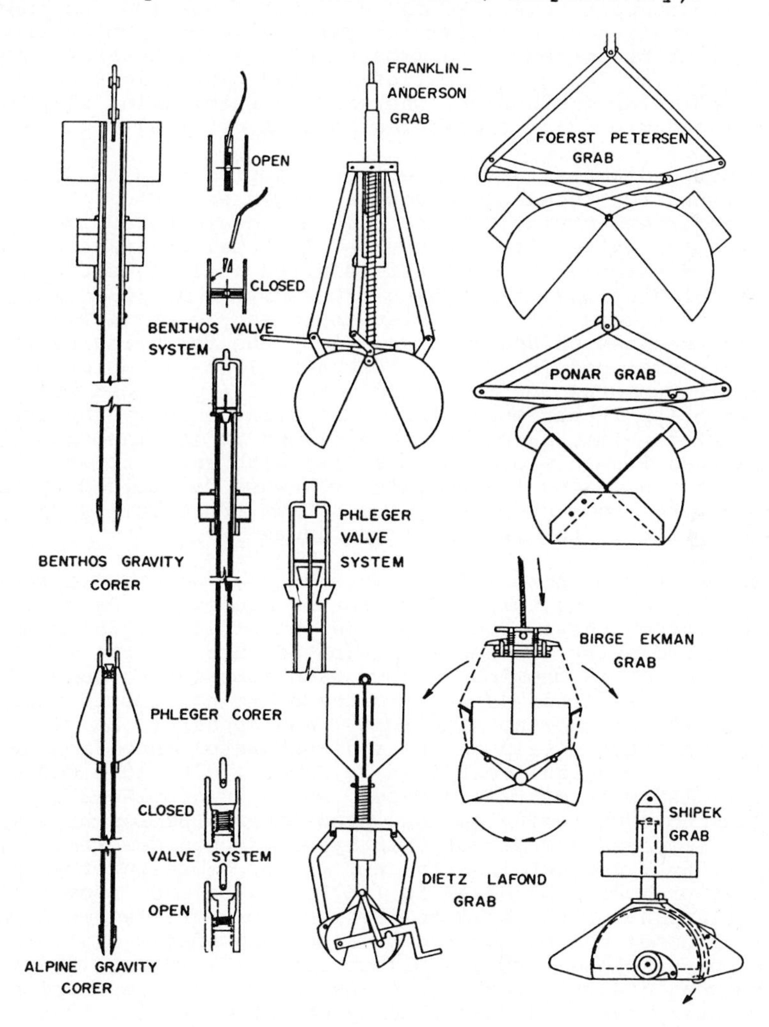

Figure 3-2 Examples of Corers and Bottom Samplers (Sly, 1969)

Sediment **sampling for "survey activities"**, e.g., for calculation of mass balances and for the study of processes, requires higher level of sophistication (Jenne *et al.*, 1980):

(i) Suspended sediment. If mass balances of suspended sediment and associated pollutants are to be calculated, a representative sample must be taken by using a depth-integrating sampling procedure; for the study of trace metals, any rubber parts of the sampling equipment are replaced by silicone ones, and exposed metal coated with a suitable plastic or replaced with Teflon; for investigations on toxic organic chemicals, sampling devices have been constructed of stainless steel for the major parts. A 0.1 μm membrane will normaly separate the particulate and dissolved phases effectively enough to provide valid data for thermodynamic geochemical calculations, whereas the 0.45 μm membrane will not, mainly with respect to Al, Fe and Mn. Centrifugation can impose difficulties if indigenous organisms in natural water samples are disrupted to "leak" out bioconcentrated metals onto the bulk water sample. Preservation of samples by quick-freezing can cause "leakage" of contaminants by cellular disruption, whereas not stabilizing samples can permit continued microbial transformations of critical pollutants.

(ii) Bottom sediments. Anoxic sediment samples require different sampling preservation techniques such as oxygen exclusion. Drying and freezing (also freeze-drying) of the samples should be avoided for material designated for extraction procedures. If total analyses or strong acid digestion is planned the sediment is dried at $60^{o}C$, crushed and stored; for mass calculations, reweighing after drying at $105^{o}C$ may become necessary. For a more differentiated approach, in particular for solid speciation studies on anaerobic samples, the following pretreatment scheme was developed (Kersten & Förstner, 1987): Samples were taken immediately from the centre of the material (collected with a grab or corer) with a polyethylene spoon, filled into a polyethylene bottle up to the surface. Immediately after arriving at the laboratory, sediments were inserted into a glove box prepared with an inert argon atmosphere. Oxygen-free conditions in the glove box were maintained by purging continuously with argon under slight positive pressure. Extractants were deaerated prior to the treatment procedure.

(iii) Extraction of pore waters. Interstitial waters are recovered from sediments by leaching, centrifugation or squeezing. Oxidation must be prevented during these procedures. Watson *et al.* (1985) showed that sediments stored prior to the separation of interstitial water yield significant changes in chemical composition compared to samples processed with 24 h of collection. *In-situ* methods are considered more promising because of their inherent simplicity, and appear to be well adapted to the study of trace metals at the sediment-water interface under field conditions. A technique described by Mayer (1976) consists of a dialysis bag filled with distilled water, which is displaced into the sediment allowing equilibrium to take place over a period of some days to weeks; an improved sampler of this type has been described by Bottomley & Bayly (1984). Another in-situ sampler for close-interval pore water studies as presented by Hesslein (1976) can be made from a clear acrylic plastic panel with small compartments pre-drilled in 1-cm steps or less. This panel can be covered by a non-degradable dialysis membrane or by a polysulphone membrane filter sheet (Carignan, 1984). (A comparison between dialysis techniques and centrifugal separation for pore water recovery has been described by Carignan *et al.*, 1985.)

Generally, it has been shown that in polluted systems entropy increases and there is a concominant increase in instability in both the physical and biological context (Wood *et al.*, 1986). Many of the analytical techniques are handicapped by **disruptive preparation** techniques which may alter the chemical forms of inorganic and organic components or lead to loss of analyte before analysis, e.g. freezing, lyophilization, evaporation, oxidation, changes in pH, light catalyzed reactions, reactions with the sample container, time delays before analysis with biologically active samples, sample contamination. In order to minimize these adverse effects, attention must be paid at least to extraction in close to 100% yield, and validation of analytical methodology with authentic samples in the same matrix.

The general experience that the environmental behavior and toxicity of an element can only be understood in terms of its actual molecular form led to the introduction of the term **"speciation"**, which is used in a vague manner both for the operational procedure for determining typical metal species in environmental samples and for describing the distribution and transformation of such species in various media (Leppard, 1983; Bernhard *et al.*, 1986; Landner, 1987; Patterson & Passino, 1987; Batley, 1989). Problems of "speciation" became particularly complex in **heterogenous systems**, e.g. in soils, aerosol particles and sediments; thermodynamic models may give suggestions as to the possible species to expect, but due to the important role of kinetically controlled processes in biogeochemistry, the actual speciation is often different from what can be expected.

3.3. Correction for Grain Size

A large number of sediment analyses, which have been performed for the inventory, monitoring and surveillance of pollution in aquatic systems have clearly shown that it is imperative, particularly for river sediments, to base these data on a **standardized procedure** with regard to particle size. With a typical example, taken from the study of Wilber & Hunter (1979) on metal pollution in the Saddle River, the situation can be demonstrated. Figure 3-3 compares the dependencies of zinc concentrations from the grain size in two selected samples taken upstream and downstream of an urbanized area of Lodi, New Jersey. In the less contaminated material from rive mile 16.6 a general decrease of zinc concen-

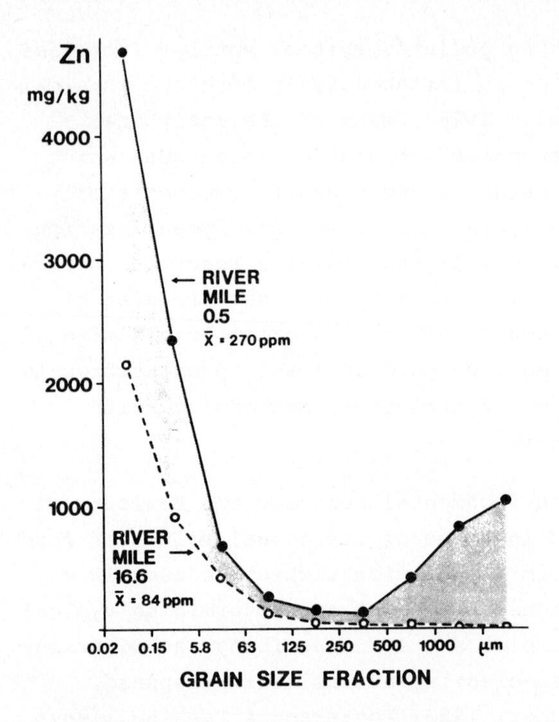

Figure 3-3

Grain Size Distribution of Zinc in Two Sediment Samples from Saddle River, Upstream (M. 16.6) and Downstream (M. 0.5) from an Urbanized Area of Lodi, New Jersey (after Data from Wilber & Hunter, 1979)

trations is found with increasing particle diameter. In the polluted material from river mile 0.5 there is a typical increase of zinc in the clay/silt fractions; a characteristic feature of the polluted sediments is the increase of zinc concentrations in the medium and coarse sand fractions. However, **selection of certain grain size fractions** - in particular for tracing of pollution sources - remains controversial for the following reasons:

(i) Since the **larger sediment fractions** are less affected by scour and transport, they may "reflect the effect of urbanization on the distribution of heavy metals over an extended period of time at a given location" (Wilber & Hunter, 1979).

(ii) The **fine sand fraction** (approx. 20 μm to 200 μm diameter) seems to be of particular interest for the differentiation of natural and pollutants metal transport, because it comprises most of the total sediment.

(iii) The **silt and clay fractions** comprise the major carriers for both natural and anthropogenic compounds (although in relatively small proportions); they are widely distributed with the sedimentation area and are least affected by grain size effects.

Different **methods** for grain size correction are compiled in Table 3-2 (after Förstner & Wittmann, 1979; De Groot *et al.*, 1982; Förstner & Schoer, 1984). These methods will mostly reduce (not elimiate!) the fraction of the sediment that is largely chemically inert, i.e. mostly the coarse-grained, feldspar and carbonate minerals, and increase the substances active in pollutant enrichment, i.e. hydrates, sulfides, amorphous and fine-grained organic materials.

Table 3-2 Methods for the Reduction of Grain Size Effects in Sediment Samples (for References see: Salomons & Förstner, 1984)

	Method	Reference (example)
I. Separation of grain size fractions (mechanical)	204 μm (sieving)	Thornton et al. (1975)
	175 μm (sieving)	Vernet and Thomas (1972)
	63 μm (sieving)	Allan (1971)
	20 μm (sieving)	Jenne et al. (1980)
	2 μm (settling tube)	Banat et al. (1972)
II. Extrapolation from regression curves	Metal/Percent 16 μm	De Groot et al. (1971)
	Metal/Percent 20 μm	Lichtfuß and Brümmer (1977)
	Metal/Percent 63 μm	Smith et al. (1973)
	Metal/specific surface area	Oliver (1973)
III. Correction for inert mineral constituents	Quartz-free sediment	Thomas (1972)
	Carbonate/quartz-free	Salomons and Mook (1977)
IV. Treatment with dilute acids or with complexing agents (determination of "mobile" fraction)	0.1 M hydrochloric acid	Gross et al. (1971)
	0.3 M HCl	Malo (1977)
	0.5 M HCl	Agemian and Chau (1976)
	25% acetic acid	Loring (1977)
	EDTA, DTPA, NTA	Gambrell et al. (1977)
V. Comparison with "conservative" elements	Metal/aluminium	Bruland et al. (1974)
	Sediment enrichment factor (Al_x/Al-backgr.)	Kemp et al. (1976)
	Rel. atomic variation	Allan and Brunskill (1977)
	M/Cs, Sc, Eu, Rb, Sm	Ackermann (1980)
	Al_x/Standard-Al	Li (1981)
	Sc_x/Standard-Sc	Schoer et al. (1982)
		Thomas and Martin (1982)

Separation of grain size is advantageous because only few samples from a particular locality are needed. However, it has been inferred that the decrease of pollutant concentrations in the medium grain size range should be even more pronounced if mechanical fractionation would more accurately separate individual particles according to their grain size. One has to consider, that "coatings", for example, of iron/manganese oxides, carbonates and organic substances on relatively inert material in respect to sorption act as substrates of pollutants in coarser grain size fractions (Förstner & Patchineelam, 1980). Nonetheless, the **fraction <63 μm** has been recommended for the following reasons (Förstner & Salomons, 1980):

- **Pollutants** have been found to be present mainly on clay/silt particles;
- this fraction is nearly equivalent to the material carried in **suspension** - the most important transport mode by far;
- **sieving** mostly does not alter pollutant concentrations (for metals even by wet sieving, when water of the same system is used);
- **numerous pollutant studies**, especially with respect to heavy metals, have already been performed on the <63 µm fraction, allowing better comparison of results.

On the other hand, it has been argued by Ackermann (1980) that separation of **fraction <20 µm**, which can also be performed with nylon sieves, should be favoured at least for coastal sediments, where the correlation with conservative elements has been found to be better with this fraction than with fraction <63 µm (see below). Also for organic pollutants, separation of fraction <20 µm seem to compare favourably with other grain size fractions (Hellmann, 1983).

Extrapolation techniques both for the grain size and specific surface area require a relative large number of samples (10-15). Further complicating is the fact that the calculation of the regression line is a tedious and mostly inaccurate procedure. The **quartz-correction** method involves fusion with potassium pyrosulfates which preferentially removes the layered silicates (clay), organic and inorganic carbon and sulfides with a residue made up of quartz plus feldspar and resistant heavy minerals such as zircone (Thomas et al., 1976).

Generally, five types of elements have been distinguished according to their distribution in sediment cores from Lake Erie (Kemp *et al.*, 1976): (i) **Diagenetically mobile** elements such as Fe, Mn and sulfur; (ii) **carbonate** elements, carbonate-C and calcium; (iii) **nutrient** elements, organic C, N, and P; (iv) **enriched** elements, such as Cu, Cd, Zn, Pb, and Hg; and (v) **conservative** elements, e.g. Si, K, Ti, Na, and Mg. Comparison of group (iv) elements of environment concern with "conservative" elements (v) seem to be particularly useful for the reduction of grain size effects, since no separation step is required.

Suitability of **reference elements** have been tested by Ackermann (1980) on sediment samples from the Ems River estuary in Northern Germany. Table 3-3 summarizes the correlation coefficient "r" between the contents of conservative elements and the percentage of grain size fraction <20 µm and <63 µm, respectively, and the slope of the regression line, the quotient s(100%)/s(0%) from the ordinate values extrapolated

Table 3-3 Correlation Coefficient, r, and Concentration Ratios, s(100%)/s(0%), for Some Potential Reference Elements (Ackermann, 1980)

	Cs	Sc	Fe	Rb	Eu	Th	Sm
Fraction < 20 μm							
r	0.987	0.982	0.858	0.958	0.945	0.32	0.878
s(100%)/s(0%)	**14.0**	**7.3**	**6.4**	**3.4**	**3.1**	**3.2**	**3.1**
Fraction < 60 μm							
r	0.919	0.937	0.789	0.900	0.947	0.944	0.911
s(100%)/s(0%)	**>20**	**15**	**9.0**	**3.7**	**3.8**	**3.8**	**3.9**

for 100% and 0% of the two grain size fractions (Figure 3-4A). According to Table 3-3, cesium appears to be the preferred reference element for two reasons: it is particularly well correlated (r = 0.987) with the percentage of the <20 μm fraction and s(100%)/s(0%) is greater than for the other elements. A test for the **applicability** of the correction procedure was made on a 140 cm long sediment core from the brackish water zone of the Ems River estuary near Emden (Figure 3-4B): The curve of bulk Zn-concentrations (broken line) in the samples vary with depth and do not give an indication of the zinc pollution, whereas the zinc data corrected for grain size (solid line) clearly indicate the very significant increase of zinc pollution during the last 100 years (dating performed with ^{137}Cs-activity).

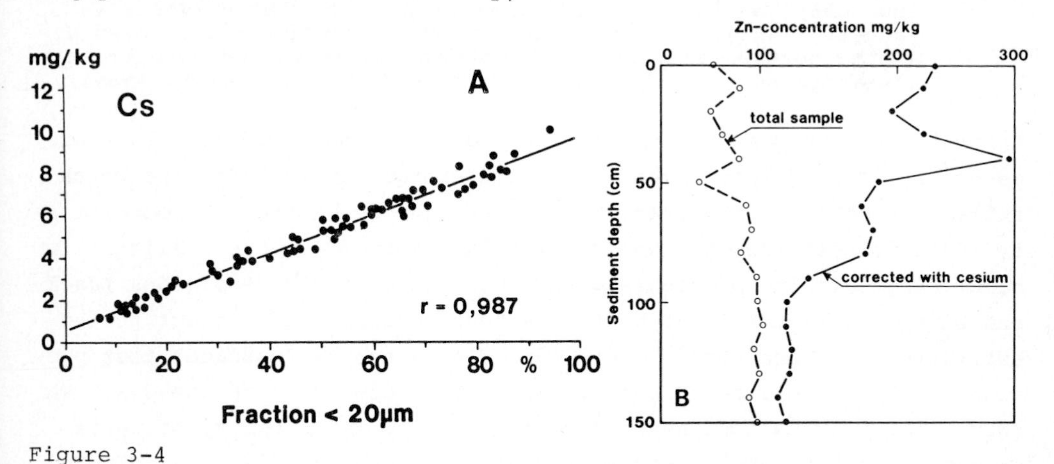

Figure 3-4

Correction of Grain Size by Conservative Element Content in Relation to Cesium (Ackermann, 1980). A: Regression Line and Correlation to Percentage of Fraction <20 μM in Sediment Samples from Ems Estuary. B: Correction of Zinc Contents in a Sediment Core from Ems Estuary.

3.4. Assessment of Critical Pools of Pollutants in Sediments

Since adsorption of pollutants onto airborne and waterborne particles is a primary factor in determining the transport, deposition, reactivity, and potential toxicity of these materials, analytical methods should be related to the chemistry of the **particle's surface** and/or to the metal species highly enriched on the surface. Basically there are three methodological concepts for determining the distribution of an element within or among small particles (Keyser *et al.*, 1978; Förstner, 1985):

- **Analysis of single particles** by X-ray fluorescence using either a scanning electron microscope (SEM) or an electron microprobe can identify differences in the matrix composition between individual particles. The total concentration of the element can be determined as a function of particle size. Other physical fractionation and preconcentration methods include density and magnetic separations.
- The **surface of the particles** can be studied directly by the use of electron microprobe X-ray emission spectrometry (EMP), electron spectroscopy for chemical analysis (ESCA), Auger electron spectroscopy (AES), and secondary ion-mass spectrometry. Depth-profile analysis determines the variation of chemical composition below the original surface.
- **Solvent leaching** - apart from the characterization of the reactivity of specific metals - can provide information on the behaviour of pollutants under typical environmental conditions. Common single reagent leachate tests, e.g. U.S. EPA, ASTM, IAEA and ICES use either distilled water or acetic acid (Theis & Padgett, 1983). A large number of test procedures have been designed particularly for soil studies; these partly used organic chelators such as EDTA and DTPA (Sauerbeck & Styperek, 1985).

Laboratory techniques for generating leachate from solid materials are generally grouped into batch and column extraction methods. The **batch extraction** methods offers advantages through its greater reproducibility and simplistic design, while the **column method** is more realistic in simulating leaching processes which occur under field conditions (Jackson *et al.*, 1984). For batch studies best results with respect to the estimation of short-term effects can be attained by **"cascade" test** procedures at variable solid/solution ratios: A procedure of the U.S. EPA (Ham et al., 1979) designed for studies on the leachability of waste products consists of a mixture of sodium acetate, acetic acid, glycine, pyrogallol, and iron sulfate. The standard leaching test developed by the Netherland Energy Research Centre (Van der Sloot *et al.*, 1984) for studies on combustion residues combines batch cascade and column procedures; the test column is filled with the material under investigation and percolated by acidulated demineralized water (pH = 4; for evaluating most relevant effects of acid precipitation) to assess short- and medium-term leaching (<50 years). In the cascade test the same quantity of material is extracted several times with fresh demineralized water (pH = 4) to get an impression of long-term leaching behavior (50-500 years).

The **soil-column techniques** fall into four major categories based on the solvent involved, namely (Fuller & Warrick, 1985):

Category A: **Dilute aqueous** inorganic and aqueous inorganic/organic - representing dilute aqueous waste streams of industrial origin containing heavy and/or toxic metals (usually dilute acidic or dilute basic aqueous solutions), municipal solid waste leachates containing both heavy and toxic metals in dilute acidic or basic solutions with soluble organic constituents, or acid rains.

Category B: **Aqueous organic** - representing solutions containing soluble organic, many of which are solvents dissolved in water from either the disposal waste stream or soil solution or both.

Category C: **Strong aqueous** acids, bases, and oxidizing agents - represented by strong concentrations of acids, bases, and oxidizing agents.

Category D: **Organic solvents** - represented by a transport system which is wholly and completely organic. The disposal solvents can contain solutes and thereby act as transport system for a host of hazardous constituents. This is particularly apparent for polar fluids such as kerosene, xylene, ethylene glycole, and isoparopol alcohol. **Special equipment** must be used to avoid health hazards from fumes.

The soil-column technique requires that only the readily measurable **properties** of the disposal system be determined. These include (1) soils data (physical and chemical), (2) transport fluid data (e.g., total organic carbon, soluble common salts, pH, etc.), (3) soil-column data, and - in particular - (4) breakthrough curves and statistical analyses (Fuller & Warrick, 1985).

One of the potential advantages of **leaching methods** are seen in obtaining relevant information from a small number of samples, one possibly being sufficient. A number of test protocols in **soil science** have been designed initially for the assessment of plant-available soil nutrients and speciation of trace metals in sewage sludge-amended soils (Jackson, 1958). In the **sediment-petrographic field** interest was focussed initially on the differentiation between authigenic and detrital phases in

Fe/Mn-concretions from deep-sea deposits (e.g. Chester & Hughes, 1967). According to Horowitz (1984) two approaches are used in chemical partitioning of sediments: The first is to determine *how* metals are retained on or by sediments - the so-called **mechanistic approach;** the second determines *where* inorganic constituents are retained on or by sediments (phase or site) - the so-called **phase approach**. Recent developments in solid phases differentiations were mainly promoted by **environmental studies**, both in soil science and in water research. For the estimation of relative bonding strength of metals in different phases, extraction procedures have been applied, both as single leaching steps and combined in sequential extraction schemes. A schematic representation of the ability of different extractant solutions to release metal ions from particulate matter is given by Pickering (1981; Figure 3-5).

EXTRACTANT TYPE	RETENTION MODE						
	Ion Exchange Sites	Surface adsorption	Precipitated (CO_3, S, OH)	Co.ppted. (amorphous hydrous oxides)	Co.ordinated to organics	Occluded (crystalline hydrous oxides)	Lattice component (mineral)
Electrolyte	$MgCl_2$						
Acetic Acid (buffer)	HOAc	HOAc/OAc^-					
(reducing)	HOAc +	NH_2OH					
Oxalic Acid (buffer)	HOx +	NH_4Ox				Light (UV)	
dil. Acid (cold)		0.4 m	HCl				
Acid (hot)	HCl +	HNO_3;	HNO_3 +	$HClO_4$			
Mixtures (+HF)		HCl +	HNO_3 +	HF			
Chelating Agents	EDTA,	DTPA					
	$Na_4P_2O_7$						
	$Na_4P_2O_7$	$+Na_2S_2O_7$					
	$Na_2S_2O_7$	+ citrate +	HCO_3^-				
Basic Solns			(alk.ppte)		NaOH		
					NaF		
Fusion (+Acid leach)		Na_2CO_3					

Figure 3-5 Schematic Representation of the Ability of Different Extractant Solutions to Release Metal Ions Retained in Different Modes or Associated with Specific Soil and Sediment Fractions (Pickering, 1981). Dashed Segments Indicate Areas of Uncertainty.

In connection with the problems arising from the disposal of solid wastes, particularly of dredged materials, **extraction sequences** have been applied which are designed to differentiate between the exchangeable, carbonatic, reducible (hydrous Fe/Mn oxides), oxidizable (sulfides and organic phases) and residual fractions (Engler *et al.*, 1977). One of the more widely applied extraction sequences of Tessier and coworkers (1979) has been modified by various authors; a version of Kersten & Förstner (1986) differentiates easily and moderately reducible components (Table 3-4).

Table 3-4 Sequential Extraction Scheme for Partitioning Sediment Samples (Kersten & Förstner, 1986)

Fraction	Extractant	Extracted Component
Exchangeable	1 *M* NH_4OAc, pH 7	Exchangeable ions
Carbonatic	1 *M* NaOAc, pH 5 w/ HOAc	Carbonates
Easily reducible	0.01 *M* NH_2OH HCl w/ 0.01 *M* HNO_3	Mn-oxides
Moderately reduc.	0.1 *M* oxalate buffer pH 3	Amorphous Fe-oxides
Sulfidic/organic	30% H_2H_2 w/ 0.02 HNO_3 pH 2 extr. w/ 1 *M* NH_4OAc-6% HNO_3	Sulfides together with organic matter
Residual	hot HNO_3 conc.	lithogenic crystall.

Despite of clear advantages of a differentiated analysis over investigations of total sample - sequential chemical extraction is probably the most useful tool for predicting long-term adverse effects from contamined solid material - it has become obvious that there are many **problems** associated with these procedures (e.g., Kersten & Förstner, 1986; Rapin et al. 1986):

(a) Reactions are **not selective** and are influenced by the duration of the experiment and by the ratio of solid matter to volume of extractants. A too high solid content, together with an increased buffer capacity may cause the system to overload; such an effect is reflected, for example, by changes of pH-values in time-dependent tests.

(b) Labile phases could be **transformed** during sample preparation, which can occur especially for samples from reducing sites.

In this respect, earlier warnings have been made by various authors, not to forget changes of the sample matrix during **recovery** and **treatment** of the material. The first indicates, that even oxic materials are not safe for changes during treatment (Thomson *et al.* 1980). The second relates to the **anoxic sediment** material, where changes are quite obvious: "The integrity of the samples must be maintained throughout manipulation and extraction" (Engler *et al.* 1977). Although these problems, particularly for anoxic sediments, have been well known for many years, they now become fully evident in the context of process studies, which will be treated in Chapter 5.

Single-extractant procedures are restricted with regard to prediction of **long-term effects** in waste deposits, e.g., of highly contaminated dredged materials, since these concepts neither involve mechanistic nor kinetic considerations and therefore do not allow calculations of release-periods. This lack can be avoided by controlled significative intensivation of the relevant parameters pH-value, redox potential and temperature combined with an extrapolation on the potentially mobilizable "pools", which are estimated from sequential chemical extraction before and after treatment of the solid material. An experimental scheme, which has originally been used by Patrick et al.(1973) and Herms & Brümmer (1978) for the study of soil suspensions and municipal waste materials, was modified by inclusion of an **ion-exchanger system** for extracting the metals released within a certain period of time each (Figure 3-6; Schoer and Förstner, 1987). For an assessment on metal oxide residues solutions were adjusted to combinations of pH 5/8 and redox 0/400 mV, circulating with 2 litres per day through columns, which contained 1:4 mixtures of waste material with quartz sand, the latter component to improve permeability; ion-exchanger resins were renewed after one week time each. In particular for the elements, of which the endpoint of release cannot be estimated from the respective cumulative curves of the water concentrations, **extrapolations** from sequential extraction data on the solid material are needed. Taking the example of zinc in Figure 3-7 the more labile "exchangeable" fractions should be released at first ("phase 1"), whereas during "release phase 2" - which is much slower than initial mobilization of acetate-extractable zinc - part of oxalate reducible compounds are dissolved. The system can be modified for different intensities of contact between solid materials and solution, by using shakers (e.g., erosion of the depot by rivers) or dialysis bags (flow-by conditions).

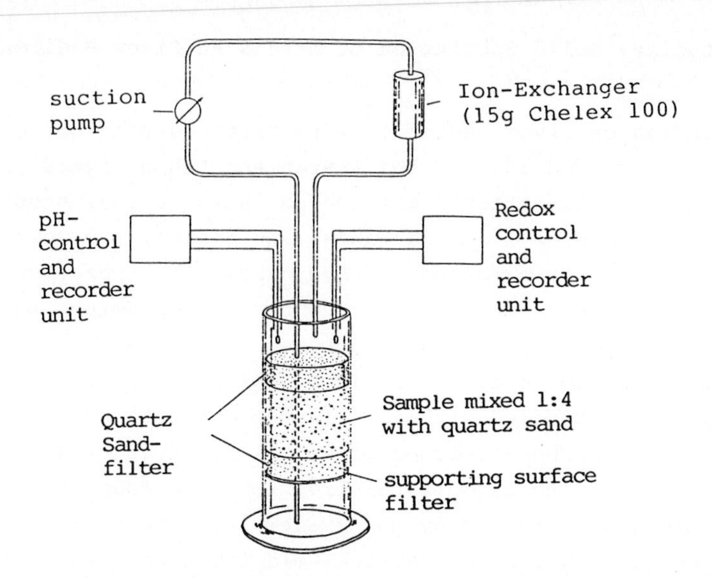

Figure 3-6 Experimental Design for Long-Term Prognosis of Metal Release from Contaminated Solids (Schoer & Förstner, 1987)

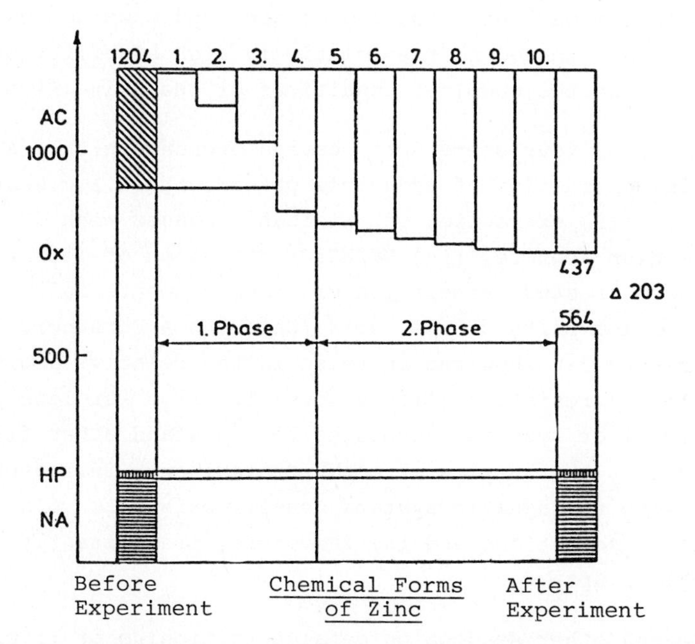

Figure 3-7 Comparison of Zinc-"Pools" in A Sample of Heat Processing Residues Before and After Treatment with pH 5/400 mV Solutions (AC = Ammonium Acetate; OX = Oxalate Buffer Solution; HP = Peroxide-Extraction; NA = Residual Fraction)

3.5. Case Studies: Solid Speciation of Metals in River Sediments

Partition studies on river sediments were first carried out by Gibbs (1973) in the suspended load of the **Amazon and Yukon Rivers**, which are less affected by civilizational influences, using a four-step leaching sequence. In the case of iron, manganese and nickel, the most significant bonding occurs, as expected, in hydroxide "coatings", whereas this type of bonding is only secondary for copper and chromium. As the hydroxide bonding decreases, a strong increase in lithogenic ("crystalline") bonding forms can be observed. In the highly polluted **Rhine River** sediments have been studied with a five-step extraction sequence (including a step for carbonate-bound elements) by Förstner & Patchineelam (1980). High percentages of detrital fractions are found for those elements which are less affected by man's activities, e.g. iron, nickel, and cobalt; on the other hand, lattice-held fractions were very low in metals such as lead, zinc, and cadmium. Lead, copper and chromium are particularly associated with the hydroxide phases. These findings can be attributed to the specific sorption of lead and copper to Fe-oxide, and by the lack of carbonate phases of chromium in natural aquatic systems. The preferential carbonate bonding of zinc and cadmium, on the other hand, can be attributed to the relatively high stability of Zn- and Cd-carbonate under the chemical conditions of the Rhine River.

A scheme consisting of **four steps** - (i) cation exchange with 1 *M* ammonium acetete, (ii) extraction of reducible phases with 0.1 *M* hydroxylamine-HCl (pH 2), (iii) extraction of oxidizable phases with 30% hydrogen peroxide/ammonium acetate, (iv) $HF/HClO_4$-digestion of detrital minerals - was used to study speciation of trace elements in river sediments from different parts of the world (Salomons & Förstner, 1980). The results in Figure 3-8 show the increase in the relative amount of metals present in the resistant (lithogenous) fraction for less polluted or unpolluted river systems. According to these and other findings, it can be argued that the surplus of metal contaminants introduced anthropogenically into the aquatic systems usually exists in relatively unstable chemical associations and is, therefore, predominantly available for biological uptake.

Apart from these and other studies on metal partitioning of river sediments, which are aimed for the identification of the **major binding sites** of trace elements, sequential extraction procedures on river sediments has been applied for the following **objectives:**

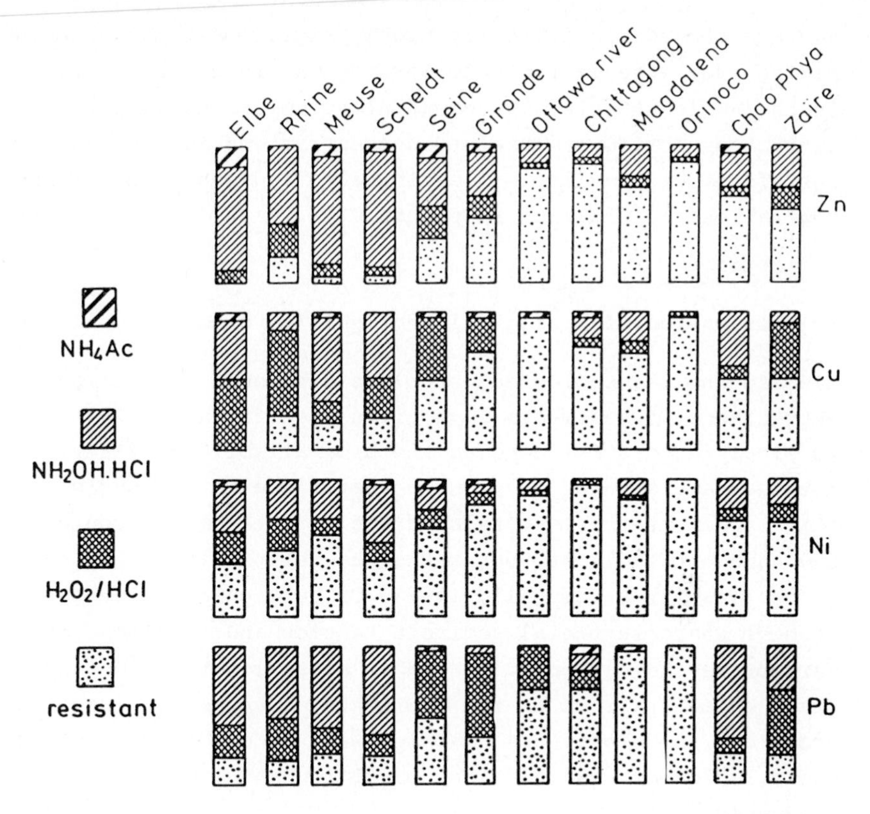

Figure 3-8 Chemical Partition of Trace Metals in Some River Sediments (after Salomons & Förstner, 1980)

(a) assessment of sources by characterization of typical inputs

In **geochemical exploration** the major applications of selective extraction procedures are (Chao, 1984): (i) to elucidate the mode of occurrence of trace metals in soils and sediments; (ii) to enhance the "geochemical contrast" between mineralized and background areas above what can be accomplished by the bulk analysis; and (iii) to differentiate between effects on metal distribution caused by anomalous mineralization, and those resulting from lithological and environmental factors. There are as yet only few studies in this field. An example has been given by Schoer (1984) with a regional survey on the distribution of thallium in soils in an area, where two point sources were effective: One was an abandoned lead-zinc mine, the other was the chimney of a cement factory which had used sulfidic roasting residues as additives to special cement. Results from leaching experiments with ammonium acetate showed statistically highly significant differences. In the mining area, the extractable portion was in the range of 4%, whereas in the soils affected by cement plant emissions approx. 18% could be extracted. The lower absolute concentrations in the latter area were therefore more available, leading also in some cases to increased concentrations of thallium in plants.

(b) estimation of biological availability of metal pollutants

Despite the methodological problems discussed in section 3.4, partial extractions of sediments have given significant insight into physico-chemical factors influencing the bioavailability of particulate trace metals. Many studies have shown that trace metal levels in various benthic organisms are best related not to total metal concentrations in the adjacent sediment, but rather to relatively easily extracted fractions (Tessier & Campbell, 1987). Of particular relevance are those studies where surficial sediment samples and specimens of **benthic invertebrates** have been collected at sites located along gradients in metal concentrations in the sediments, such as for longitudinal river profiles. In Chapter 5.4 examples will be given and factors influencing the relation between sediment partition and biological uptake will be discussed.

(c) evaluation of diagenetic effects

The typical effects of the earliest stages of "diagenesis" (involving transformations of organic matter, "aging" of mineral components and formation of new equilibria between solid and dissolved species) have been demonstrated by Salomons (1980) with respect to the behaviour of trace metals at the sediment/seawater interface (Table 3-5).

Table 3-5 Percentage of Cadmium and Zinc not Released from River Suspended Matter after Treatment with NaCl-Solution (Salomons, 1980)

Adsorption time:	1 day	3 days	8 days	24 days	60 days
Cadmium	24%	30%	33%	37%	40%
Zinc	60%	67%	70%	74%	88%

Desorption was studied by adding cadmium and zinc to suspended matter in river water; after adsorbing periods of 1, 3, 8, 24 and 60 days, NaCl was added to the suspension to increase the chloride concentration to seawater concentration. After an adsorption period of only one day, 24% of the adsorbed cadmium and 60% of the adsorbed zinc remains bound to the sediment; after 60 days 40% of the cadmium and 88% of the zinc bound to the sediment is not released after NaCl treatment. An extrapolation can be made to **geologic time scales** by a comparison of the bonding intensity of stable metal isotopes and their unstable counterparts - the latter supplied from radioactive emissions of nuclear power and reprocessing plants. In Figure 3-9 the effects are shown of sequential leaching of a sediment sample from the lower Rhône River in France (Förstner & Schoer, 1984). The reducing agents hydroxylamine (pH 2) and oxalate buffer (pH 3) only extract 15% of the natural stable manganese while the artificial isotope Mn-54 from the reprocessing plant is mobilized at more than 80% by these treatments. One of the implications of these data is that under reducing conditions there should be a more intensive release of the radioactive manganese nuclides compared to their stable counterparts, i.e. the "availability" of the geologically younger Mn-nuclide for short-term chemical interaction and biological uptake is much higher than for the stable, "geochemical" Mn-isotopes.

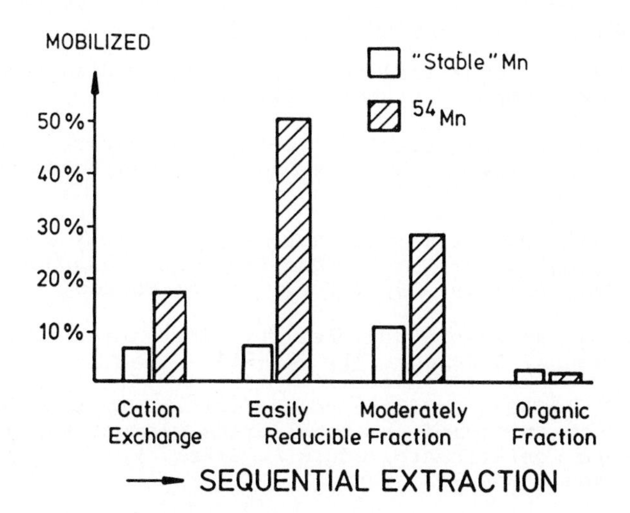

Figure 3-9 Comparison of Chemical Extractability of Artificial and Stable Isotopes of Manganese from a Sediment Sample of the Rhône River (Förstner & Schoer, 1984)

References to 3.1 "Objectives"

Anonymus (1978a) *Water Quality Survey. A Guide for the Collection and Interpretation of Water Quality Data*. Studies and Reports in Hydrology No. 23, 350 p. Paris: Unesco.

Anonymus (1978b) *GEMS: Global Environmental Monitoring System*. 313 p. Geneva: World Health Organization.

Bear, F.E. (Ed.) (1964) *Chemistry of the Soil*. 2nd ed. New York: Reinhold Publ. Co.

Golterman, H.L., Sly, P.G. & Thomas, R.L. (1983) *Study of the Relationship between Water Quality and Sediment Transport*. Technical Papers in Hydrology No. 26, 231 p. Paris: Unesco

Levinson, A.A. (1974) *Introduction to Exploration Geochemistry*. Calgary: Applied Publishing Ltd.

References to 3.2 "Sampling"

Bernhard, M., Brinckman, F.E. & Sadler, P.S. (Eds.)(1986) *The Importance of Chemical "Speciation" in Environmental Processes*. Dahlem-Konferenzen, Life Sciences Research Report 33, 763 p. Berlin: Springer-Verlag.

Bottomley, E.Z. & Bayly, I.L. (1984) A sediment pore water sampler used in root zone studies of the submerged macrophyte, Myriophyllum spicatum. *Limnol. Oceanogr.* 29, 671-673.

Carignan, R. (1984) Interstitial water sampling by dialysis. Methodological notes. *Limnol. Oceanogr.* 29, 667-670.

Carignan, R., Rapin, F. & Tessier, A. (1985) Sediment pore water sampling for metal analysis: A comparison of techniques. *Geochim. Cosmochim. Acta* 49, 2493-2497.

Golterman, H.L., Sly, P.G. & Thomas, R.L. (1983) *Study of the Relationship between Water Quality and Sediment Transport*. Technical Papers in Hydrology No. 26, 231 p. Paris: Unesco

Hesslein, R. (1976) An in-situ sampler for close interval pore water studies. *Limnol. Oceanogr.* 21, 912-914.

Jenne, E.A. *et al.* (1980) Sediment collection and processing for selective extraction and for total trace element analysis. In *Sediments and Contaminants*, ed. R.A. Baker. Vol. 2, pp. 169-191. Ann Arbor: Ann Arbor Sci. Publ.

Kersten, M. & Förstner, U. (1987) Effect of sample pretreatment on the reliability of solid speciation data of heavy metals - implication for the study of diagenetic processes. *Mar. Chem.* 22, 299-312.

Landner, L. (Ed.)(1987) *Speciation of Metals in Water, Sediment and Soil Systems*. Lecture Notes in Earth Sciences No. 11, 190 p. Berlin: Springer-Verlag.

Leppard, G.G. (Ed.)(1983) *Trace Element Speciation in Surface Waters and its Ecological Implications*. Proc. NATO Advanced Research Workshop, Nervi/Italy, Nov. 2-4, 1981. 320 p. New York: Plenum Press.

Mayer, L.M. (1976) Chemical water sampling in lakes and sediments with dialysis bags. *Limnol. Oceanogr.* 21, 909-911.

Patterson, J.W. & Passino, R. (Eds.)(1987) *Metals Speciation, Separation, and Recovery*. 779 p. Chelsea/MI: Lewis Publ.

Sly, P.G. (1969) Bottom sediment sampling. *Intern. Assoc. Great Lakes Res.*, Proc. *12th Conf. Great Lakes Res.*, pp. 230-239

Watson, P.G., Frickers, P.E. & Goodchild, C.M. (1985) Spatial and seasonal variations in the chemistry of sediment interstitial waters in the Tamar estuary. *Estuar. Coast. Shelf Sci.* 21, 105-119.

Wood, J.M. *et al.* (1986) Speciation in systems under stress (group report). In: *The Importance of Chemical "Speciation" in Environmental Processes*, eds. M. Bernhard *et al.*, Dahlem Konferenzen, pp. 425-441. Berlin: Springer-Verlag

References to 3.3 "Grain Size Corrections"

Ackermann (1980) A procedure for correcting the grain size effect in heavy metal analysis of estuarine and coastal sediments. *Environ. Technol. Lett.* 1, 518-527.

De Groot, A.J., Zschuppe, K.H. & Salomons, W. (1982) Standardization of methods of analysis for heavy metals in sediments. In *Sediment/-Freshwater Interaction*, ed. P.G. Sly. *Hydrobiologia* 92, 689-695.

Förstner, U. & Patchineelam, S.R. (1980) Chemical associations of heavy metals in polluted sediments from the lower Rhine River. In *Particulates in Water*, eds. M.C. Kavaunagh & J.O. Leckie. *Adv. Chem. Ser. Amer. Chem. Soc.* 189, 177-193.

Förstner, U. & Salomons, W. (1980) Trace metal analysis on polluted sediments. I. Assessment of sources and intensities. *Environ. Technol. Lett.* 1, 494-505.

Förstner, U. & Wittmann, G. (1979) *Metal Pollution in the Aquatic Environment*. Berlin: Springer-Verlag.

Hellmann, H. (1983) Korngrößenverteilung und organische Spurenstoffe in Gewässersedimenten und Böden. *Fresenius Z. Anal. Chem.* 316, 286-289.

Kemp, A.L.W. *et al.* (1976) Cultural impact on the geochemistry of sediments in Lake Erie. *J. Fish. Res. Board Can.* 33, 440-462.

Salomons, W. & Förstner, U. (1984) *Metals in the Hydrocycle.* Berlin: Springer-Verlag.

Thomas, R.L. *et al.* (1976) Surficial sediments of Lake Erie. *J. Fish. Res. Board Can.* 33, 385-403.

Wilber, W.G. & Hunter, J.V. (1979) The impact of urbanization on distribution of heavy metals in bottom sediments of the Saddle River. *Water Resour. Bull.* 15, 790-800.

References to 3.4 "Assessment of Critical Pools of Pollutants"

Chester, R. & Hughes, M.J. (1967) A chemical technique for the separation of ferromanganese minerals, carbonate minerals and adsorbed trace elements from pelagic sediments. *Chem. Geol.* 2, 249-262.

Engler, R.M. *et al.* (1977) A practical selective extraction procedure for sediment characterization. In *Chemistry of Marine Sediments*, ed. T.F. Yen, pp. 163-171. Ann Arbor: Ann Arbor Sci Publ.

Förstner, U. (1985) Chemical forms and reactivities of metals in sediments. In *Chemical Methods for Assessing Bio-Available Metals in Sludges and Soils*, eds. R. Leschber *et al.*, pp. 1-30. London: Elsevier Applied Sci.

Fuller, W.H. & Warrick, A.W. (1985) *Soils in Waste Treatment and Utilization.* Boca Raton/Fla.: CRC Press

Ham, R.K. *et al.* (1979) *Background Study on the Development of a Standard Leaching Test.* U.S. EPA-600/2-79-109. Cincinnati: U.S. Environmental Protection Agency.

Herms, U. & Brümmer, G. (1978) Löslichkeit von Schwermetallen in Siedlungsabfällen und Böden in Abhängigkeit von pH-Wert, Redoxbedingungen und Stoffbestand. *Mitt. Deutsche Bodenkundl. Ges.* 27, 23-34.

Horowitz, A.J. (1984) *A Primer on Trace Metal-Sediment Chemistry.* Open-File Rep 84-709, 82 p. Doraville/GA: U.S. Geological Survey

Jackson, D.R., Garrett, B.J. & Bishop, T.A. (1984) Comparison of batch and column methods for assessing leachability of hazardous waste. *Environ. Sci. Technol.* 18, 668-673.

Jackson, M.L. (1958) Soil Chemical Analysis. Englewood Cliffs/NJ: Prentice-Hall

Kersten, M. & Förstner, U. (1986) Chemical fractionation of heavy metals in anoxic estuarine and coastal sediments. *Water Sci. Technol.* 18, 121-130.

Keyser, T.R. *et al.* (1978) Characterizing the surface of environmental particles. *Environ. Sci. Technol.* 12, 768-773.

Patrick, W.H., Williams, B.G. & Moraghan, J.T. (1973) A simple system for controlling redox potential and pH in soil suspensions. *Soil Sci. Soc. Amer. Proc.* 37, 331-332.

Pickering, W.F. (1981) Selective chemical extraction of soil components and bound metal species. *CRC Crit. Rev. Anal. Chem. Nov.*, pp. 233-266.

Rapin, F. *et al.* (1986) Potential artifacts in the determination of metal partitioning in sediments by a sequential extraction procedure. *Environ. Sci. Technol.* 20, 836-840.

Sauerbeck, D. & Styperek, P. (1985) Evaluation of chemical methods for assessing the Cd and Zn availability from different soils and sources. In *Chemical Methods for Assessing Bio-Available Metals in Sludges and Soils*, eds. R. Leschber *et al.*, pp. 49-66. London: Elsevier Applied Sci.

Schoer, J. & Förstner, U. (1987) Abschätzung der Langzeitbelastung von Grundwasser durch die Ablagerung metallhaltiger Feststoffe. *Vom Wasser* 69, 23-32.

Tessier, A., Campbell P.G.C. & Bisson, M. (1979) Sequential extraction procedure for the speciation of particulate trace metals. *Anal. Chem.* 51, 844-851.

Theis, T.L. & Padgett, L.E. (1983) Factors affecting the release of trace metals from municipal sludge ashes. *J. Water Pollut. Control Fed.* 55, 1271-1279

Thomson, E.A. *et al.* (1980) The effect of sample storage on the extraction of Cu, Zn, Fe, Mn, and organic material from oxidized estuarine sediments. *Water Air Soil Pollut.* 74, 215-233.

Van der Sloot, H.A., Piepers, O. & Kok, A. (1984) *A Standard Leaching Test for Combustion Residues*. Shell BEOP-31. Petten/Netherlands: Studiegroep Ontwikkeling Standaard Uitloogtesten Verbrandingsresiduen

References to 3.5 "Solid Speciation of Metals in River Sediments"

Chao, T.T. (1984) Use of partial dissolution techniques in geochemical exploration. *J. Geochem. Explor.* 20, 101-135.

Förstner, U. & Schoer, J. (1984) Some typical examples of the importance of the role of sediments in the propagation and accumulation of pollutants. In *Sediments and Pollution in Waterways - General Considerations*. IAEA-TecDoc-302, pp. 137-158. Vienna: International Atomic Energy Agency.

Gibbs, R.J. (1973) Mechanisms of trace metal transport in rivers. *Science* 180, 71-73.

Salomons, W. (1980) Adsorption processes and hydrodynamic conditions in estuaries. *Environ. Technol. Lett.* 1, 356-365.

Salomons, W. & Förstner, U. (1980) Trace metal analysis on polluted sediments. II. Evaluation of environmental impact. *Environ. Technol. Lett.* 1, 506-517.

Schoer, J. (1984) Thallium. In *Handbook of Environmental Chemistry*, ed. O. Hutzinger, Vol. 3 Part C, pp. 143-214. Berlin: Springer-Verlag.

Tessier, A. & Campbell, P.G.C. (1987) Partitioning of trace metals in sediments: Relationship with bioavailability. In *Ecological Effects of* In Situ *Sediment Contaminants*, eds. R.L. Thomas *et al.*, *Hydrobiologia* 149, 43-52.

Additional Themes (Examples)

"**Sampling Design** for Specific Sedimentary Environments - Estuaries, Coastal Basins, Lakes, Groundwater, Mine Tailings, Dredged Materials"

"**Sampling Devices** and Strategies for Suspended Matter in Rivers"

"**Gravity Fractionation** (incl. Data Interpretation) of Sediment Samples"

"**Non-Destructive Analysis** of Sediment Samples" (e.g., X-Ray Methods)

"Extraction and Analysis of Persistent **Organic Chemicals** in Sediments"

"Experimental Methods for the Study of **Interfacial Processes**, both at the Surface of Sediment Particles and at In-Situ Sediment Boundaries"

4. PARTICLE ASSOCIATIONS

4.1. Particles in Water

Due to the strong interactions with particulate matter the behaviour of contaminants such as heavy metals and organic chemicals in surface waters is regulated by the transport, dispersal and sedimentation of aquatic solid materials. Chemical processes therefore are strongly influenced by the **suspended matter concentrations** and due to the shallowness of the waters by the deposited sediments. Compared with other parts of the hydrological cycle, rivers and estuaries are characterized by high suspended matter concentrations (Figure 4-1; Kranck, 1980).

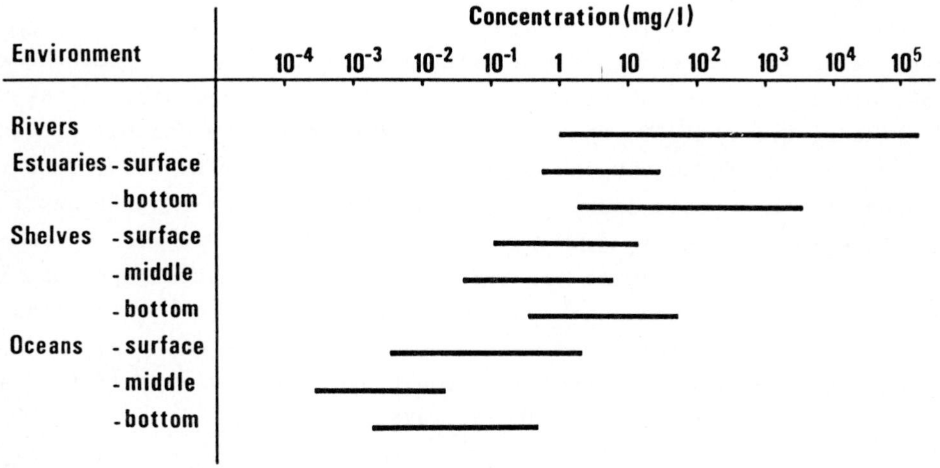

Figure 4-1. Particulate Concentrations in Different Aquatic Systems (Kranck, 1980)

The **transport of sediments** is strongly related to hydrological and geomorphological phenomena. The main processes to be considered are (Raudkivi, 1976): (i) Erosion of the sediments from a bottom or a bank; (ii) vertical transport of the particles in the body of the water in which it is carried; (iii) horizontal transport by the flow; (iv) deposition on the bottom, and (v) compaction (consolidation) of the deposit. In general, for sediment particles to start moving, the flow of water must exceed a certain critical velocity. The "Hjulström-curve" shows that for coarse particles that velocity for erosion and deposition is almost equal, whereas for the erosion of fine cohesive sediments a stronger current is needed than the threshold velocity at which these particles are deposited.

In general, increase of water velocity in **rivers** strongly enhances concentration of suspended material; a modifying factor for these response is - among other parameters such as intensity of weathering and erodibility - the availability of finer-grained material, which has been deposited during reduced water discharge (Müller & Förstner, 1968). Similarly in **lakes**, particle size fractions may be limited in their occurrence due to a scarcity in supply (Wolfe, 1964; Sly, 1978). Typical **estuarine** circulation is characterized by seaward transport of suspended particles by the river and by residual landward flow of settling particles along the bottom. The maximum concentration of suspended matter appears to occur at the tip of the saltwater wedge in the estuary, and the turbidity maximum may extent into the freshwater tidal area (Postma, 1967).

In many cases man has enhanced the **sedimentation** by creating freshwater basins, diverting river stretches and by building of harbours. It has been noted by Gross (1972) that "solid waste disposal by coastal cities modify shorelines and cover the adjacent ocean bottom with characteristic deposits on a scale large enough to be geologically significant". A typical example is given from Long Island Sound near New York where waste discharges from **dumping** are ten times higher than the input of river-borne suspended sediment. To visualize the volume of the dumping of waste into the coastal ocean near New York City, one must imagine that the annual discharge spread uniformely over Manhattan would form a layer 13 cm thick each year (Goldberg, 1976).

In the case of harbor areas the accumulated (mostly contaminated) sediments have to be removed and dumped elsewhere; this means that they may end in places they never would reach by natural processes, and in a different **geochemical environment** (Salomons, 1987; see also Chapter 7). Typical locations, where sediment-associated pollutants can be accumulated, are shown in Figure 4-2 (Shea, 1988). Of particular interest are those sites, where - due to the hydrodynamic situation - fine-grained particles are deposited, which usually contain highest concentrations of toxic chemicals (Chapter 3.3).

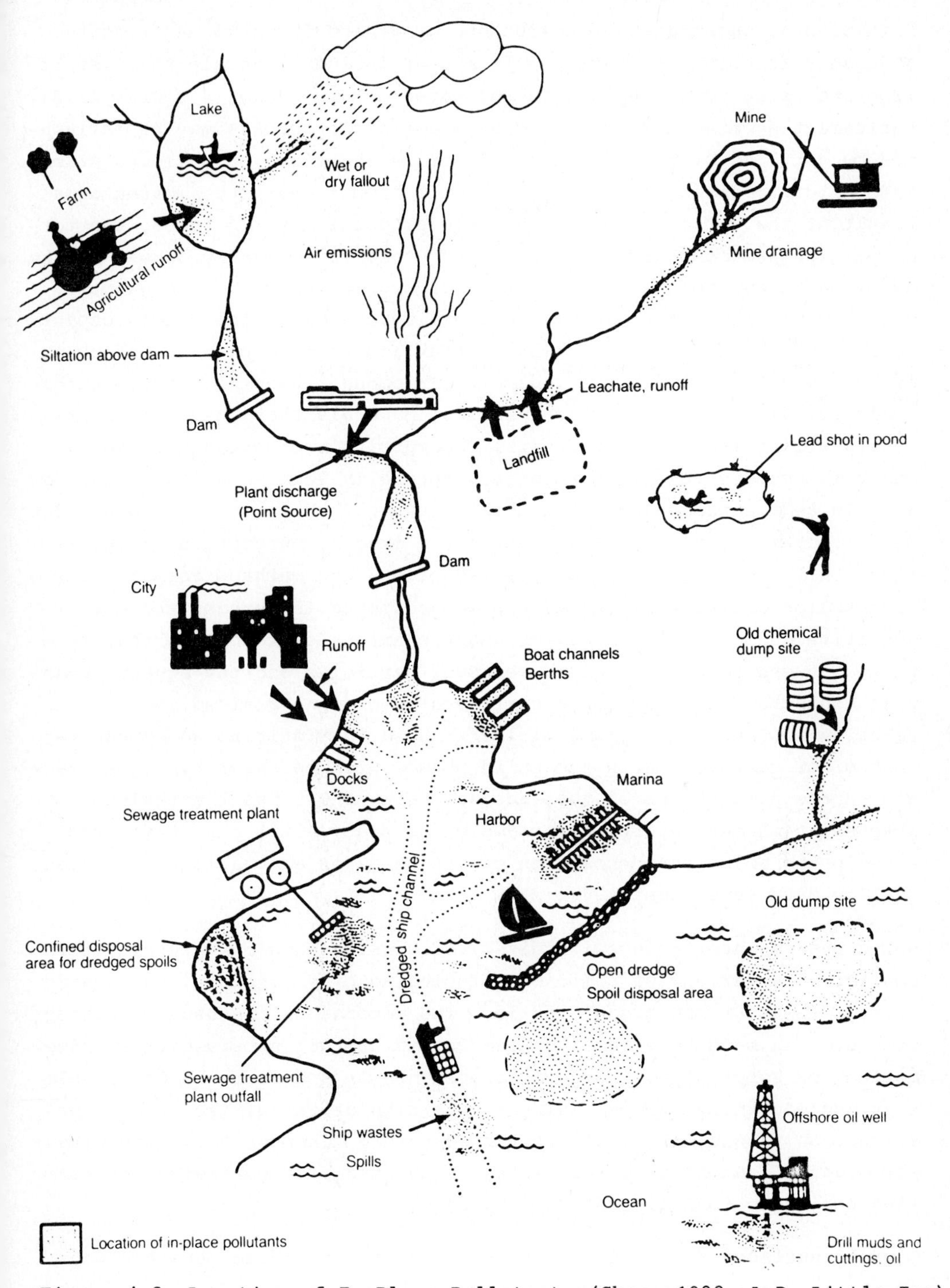

Figure 4-2 Location of In-Place Pollutants (Shea, 1988, A.D. Little Inc)

Particles as substrates of pollutants originate from two major sources: **Endogenic fractions** of particulate matter include minerals that result from processes occurring within the water column (Jones & Bowser,1978). Enrichment of minerals generated by endogenic processes may be influenced by settling of particulates, filtering organisms, and flocculation. Endogenic processes exhibit a distinct temporal character, often as a result of the variation of the organic productivity. In lakes, the total particulate concentration of trace metals is generally lowest in the hypolimnion due to the decomposition of organic matter. Consequently, net biogenic flux, for example, of metals depends on the lake's capacity to produce organic particulate matter and to decompose it before it is buried definitely in the sediment (Salomons & Baccini, 1986). **Authigenic** (or diagenetic) fractions include minerals that result from processes within deposited sediments. Decomposition of organic matter, which is mediated by microorganisms, generally follows a finite succession in sediments depending upon the nature of the oxidizing agent (see Berner, 1981); the successive events are oxygen consumption (respiration), nitrate reduction, sulfate reduction, and methane formation. The composition of **interstitial waters** in sediments is perhaps the most sensitive indicator of the types and the extent of reactions that take place between pollutant-loaded sediment particles and the aqueous phase that contacts them. The large surface area of fine-grained sediment in relation to the small volume of its trapped interstitial water ensures that minor reactions with the solid phases will be shown by major changes in the composition of the aqueous phase. Thus, the theoretical assessment of the nature, for example, of trace-metal phases via the equilibrium solution composition can be used for evaluation of sediment quality data (see Chapter 6).

Solid phases interacting with dissolved constituents in natural waters consist of a variety of components including clay minerals, carbonates, quartz, feldspar and organic solids. The components are usually coated with hydrous manganese and iron oxides and organic substances. A prime medium for sorption of inorganic components by sediments is metastable iron and manganese oxides, which have a high degree of isomorphic substituation (Jenne, 1977). As mentioned before (Chapter 2.3), the natural organic content of particulates is the decisive parameter controlling the "sorption" of organic chemicals.

Organic surfaces for metal sorption could form in three possible ways (Hart, 1982): (i) from organisms such as bacteria and algae; (ii) by the breakdown of plant and animal material and by the aggregation of lower-molecular weight organics; and (iii) by organic matter of lower-molecular weight sorbed onto clay or metal oxide substrates (Davis & Gloor, 1981). Although the difference between these three surface types is not well understood with respect to metal uptake, there is a general agreement that at least on major binding mechanism involves salicylic entities; other strong binding entities (such as peptides) may also be present in some systems. Tipping (1981) suggests that at least part of the organic matter adsorbed onto the particulate matter in natural waters has carboxylic and phenolic functional groups available for binding with trace metals. The trace metal adsorption capacity of organic matter is generally between that for metal oxides and clays.

Colloids seem to have been the "forgotten component in aquatic systems" (Allan, 1986) partly because few researchers have the facilities or the inclination to investigate these substances. Many of the natural colloidal matter in continental waters are humic acid polymers, containing functional groups such as COOH, OH, C=O, NH_2 and SH_2 (Frimmel & Christman, 1988). These colloids which are mostly negatively charged, interact with metals to form metal-organic complexes by ion exchange and chelation processes (Weber, 1988), and with organic chemicals by protonation, Van der Waals forces, coordination complexation, and hydrophobic bonding. In the latter form they are typical carriers of organic micropollutants in freshwater (Carter & Suffet, 1982), estuarine (Poirrier *et al.*, 1972), and seawater (Pierce *et al.*, 1974). Colloids, among other properties, are easily biodegradable by microbial processes and may act in the initial stage of contaminant bioaccumulation; they generally play a major - but not well understood - role in the transfer of pollutants between different compartments in the aquatic environment. Field studies by Baker *et al.* (1986) and Brownawell & Farrington (1986) indicate the possible influence of colloidal and dissolved high molecular (humic) organic material on the distribution of organic chemicals between aqueous and solid phases; in the latter study PCB's in pore waters considerably exceeded water solubilities. A three phase model including not only solid material and water-dissolved contaminant, but also non-filtrable microparticles and macromolecular organic matter indicates that colloid-associated contaminants might be the dominant species in equilibrium with the solid phase (Baker *et al.*, 1986). In laboratory experiments with isolated nonfiltrable humic material this association was quantified (Hassett & Milicic, 1985; McCarthy & Jimenez, 1985; Chiou *et al.*, 1986).

4.2. Factors Influencing Partition Coefficients

If the partition coefficients (the log of the ratio of the concentration in particulates to the concentration in solution) was dependent only on the solubility or related properties of a specific chemical, and if equilibrium was always attained between the particulate and aqueous phases, then the mass of a **chemical transported on the particulate phase** should be readily predicted (Allan, 1986; Table 4-1):

Table 4-1 Percent of Contaminant in the Particulate Fraction (If Controlled Solely by Partition Coefficient and Concentration of Suspended Solids). From Allan (1986).

Suspended Solids Concentration (mg/L)	K_D (Partition Coefficient)					
	10^7	10^6	10^5	10^4	10^3	10^2
	(Percent in Particulate Fraction)					
1000	100[a]	100	100	91	50	9
500	100	100	98	85	33	5
100	100	100	91	50	9	1
50	100	98	85	33	5	0[b]
30	100	97	75	25	3	0
10	100	91	50	10	1	0
5	98	85	33	9	0	0
1	91	50	10	5	0	0

[a] 100% means >99%; assuming than an equilibrium is established. Even for the most insoluble compounds there will always be some finite amount in solution.

[b] Zero means <1%; assuming that all natural waters contain some finite amount of suspended solids, there will always be a finite amount of partitioning.

An example has been given by Allan (1986) for the **Niagara River**, which has a mean annual suspended particulate load of 8.4 mg/L. Assuming this suspended particulate concentration, some 50% or greater of toxic chemicals and elements with partition coefficients larger than 10^5 should be transported in the particulate phase. The mass transported by the particulate phase increases with increasing particulate concentration and with increasing partition coefficient.

Several factors affect the partition coefficients of organic chemicals and elements in aquatic systems and examples of such influences are given here:

Effects of Phase Separation, Sample Preparation and Grain Size

Typical effects of the separation methods on the partition coefficient have been studies by Calmano (1979). Data from radionuclide analysis in Table 4-2 indicate that distribution coefficients between solid and aqueous phases generally are higher from **filtration** methods than from **centrifuge** separation; for cesium and chromium the difference is a factor of about 10, for iron even higher.

Table 4-2 Influence of Separation Method on the Calculation of K_D Factors for the Example of Main River Water (Calmano, 1979)

Element	K_D from Centrifugation	K_D from Filtration (0.45 m)
Arsenic	3.16×10^4	7.61×10^4
Cerium	6.62×10^4	3.27×10^5
Cesium	4.86×10^4	5.10×10^5
Chromium	4.65×10^4	5.14×10^5
Cobalt	2.74×10^4	4.07×10^4
Iron	4.63×10^4	1.90×10^6
Zinc	3.86×10^4	8.05×10^4

Other methodological problems result from sampling and sample preparation, particularly with respect to the solid phase (Chapter 3.2). The effect of preparation methods on experimental determinations of K_D-factors of radionuclides between aqueous phases and aquatic sediments has been reviewed by Duursma (1984). There are significant effects of the **drying procedures**. The influence of grain size on the distribution coefficient can be enormous, as demonstrated by the data in Table 4-3:

Table 4-3 K_D-Factors of Radionuclides in Relation to the Grain Size in a Sediment Sample from the Wadden Sea (after Duursma, 1984)

Grain Size Fraction (μm)	Weight %	**Distribution Coefficients x 10^2** Cs-137	Fe-59	Zn-65	Co-60	Zr-95	Ce-144
< 4	3.7	6.2	540	112	220	670	124
4- 8	8.4	16	510	380	430	1220	540
8-16	5.6	12	370	490	59	1040	950
16-32	9.5	5.4	53	250	65	66	120
32-64	21.7	1.6	76	-	4	42	3
> 64	51.1	-	-	-	6	1	1

Effect of Suspended Particle Concentration

Connor and Connolly (1980) were among the first to show that there is an inverse relationship between adsorbing solids and partition coefficient for organic chemicals. Using a natural sediment, a change in particle concentration from 100 mg/L to 1000 mg/L lowered the partition coefficient for **Kepone** from 5900 to 3400. Above sediment concentrations of 20 mg/L, the sorption relationship is no longer linear and progressively less of the chemical becomes adsorbed, i.e., the K_D decreases. Voice *et al.* (1983) have suggested that the diversity of solute/sorbent combinations that exhibit this behaviour tends to rule out any explanation based on specific chemical interactions and rather indicates a nonspecific, perhaps physical explanation. It is proposed that the observed change in partitioning behaviour due to solids concentration can be attributed to a transfer of sorbing, or solute-binding, material from the solid phase to the liquid phase during the course of the partitioning experiment. This material, whether dissolved, macromolecular or microparticulate in nature, is not removed from the liquid phase during the separation procedure and is capable of stabilizing the compound of interest in solution. The amount of material contributed to the liquid phase is most likely proportional to the amount of solid phase present, and thus, the capacity of the liquid phase to accomodate solute depends upon the concentration of solids in the system. No simple means exists by which the presence of these microparticles can be quantified, and thus, extrapolation of the results to other systems can only be inferred. Based on these findings, the following precautions are in order relative to the use of partitioning relationships developed by laboratory measurements (Voice *et al.*, 1983):

- Partition coefficients developed at one concentration of solids are not necessarily appropriate at other solids concentrations;
- Partition coefficients produced in studies using different techniques are not necessarily comparable.

The **solids effect** can be quantified, and a reasonable estimate of the partition coefficient can be produced on the basis of the organic carbon concent of the solid, the octanol-water partition coefficient of the solute, and the concentration of the solids. The solids effect is likely to occur in the environment although the extent of the effect relative to that observed in the laboratory is unknown. Laboratory partition coefficients are not, therefore, directly applicable to real systems.

Source of Contaminant; Residence Time

The ratios between dissolved and solid contaminant fractions are firstly influenced by the respective **inputs** and subsequently by the **interactions** taking place within the different environmental compartments. Direct emissions, for example of cadmium, into the environment from waste materials are approximately 10-fold higher from solid materials (pigments, phosphate fertilizers, sewage sludge, municipal and mining wastes, smelting residues, etc.) than from dissolved inputs (from lead-zinc mines, sewage treatment plant, effluents from battery factories, electroplating plants, etc.). This is especially valid for river systems, where equilibrium between the solution and the solid phases can often not be achieved completely due to the short **residence times**; Bowen (1977) has noted the example of the Thames River flowing 250 km in 14 hours, while chemical equilibrium would require more than 100 hours in some instances. The bulk of detrital trace-element particulates never leaves the solid phase from initial weathering to ultimate deposition. Similarly, metal dust particles (e.g., from smelters) and effluents containing metals associated with inorganic and organic matter undergo little or no change after being discharged into a river. In natural systems, such as in the Amazon River and to a large extent also in the Mississippi River, more than 90% of the metal load is transported by particulate matter (Table 4-4). A similar sequence of the ratios between particulate and dissolved heavy metals has also been found for polluted systems; typically, however, the dissolved fractions in polluted waters are significantly higher than in less polluted systems, particularly for metals such as Cd, Zn, and Cu.

Table 4-4 Particulate-Bound Metal as Percentage of Total Discharge (Förstner, 1984; after data from Gibbs, 1973; Trefry & Presley, 1976; Kopp & Kroner, 1968; Heinrichs, 1975).

	Amazon River	Mississippi River	Polluted Rivers in U.S.A.	F.R.G.
Cadmium	-	88.9	-	30
Zinc	-	90.1	40	45
Copper	93	91.6	63	55
Manganese	83	93.5	-	8 - 97
Chromium	83	98.5	76	72
Lead	-	99.2	84	79
Iron	99.4	99.9	98	98

Sorption Kinetics

Typical adsorption of metals increase from near nil to near 100% as pH increases through a critical range 1-2 units wide (Benjamin *et al.*, 1982). It is important to note, that the location of the pH-adsorption-"edge" depends on **adsorbent concentration**. This effect is due to the existence of a range of **"specific" site-binding energies**. High-energy adsorption sites, since they are fewer in number than lower energy sites, become limiting first. As lower energy sites are gradually filled, the overall binding constant decreases. There is a typical temporal evolution of the sorption processes; four different types of evolution (rapid or slower adsorption to nearly 100%; rapid or slow adsorption at a lower level) have been distinguished from experiments using radioisotopes (Schoer & Förstner, 1985). These processes are influenced by the hydrological and chemical conditions; sorption of cesium, for example, is typically lowered in the presence of Ca- and Mg-ions. In the few cases, where kinetics were investigated, surface reactions were not found to be a single step reaction (Chen *et al.*, 1973; Anderson, 1981). Experiments by Benjamin & Leckie (1981) showed a rapid and almost complete metal uptake process perhaps lasting no more than one hour, followed by a second, slower uptake process perhaps lasting days, or possible months; the first effect was thought to be true adsorption, and the second to be slow adsorbate diffusion into the solid substrate.

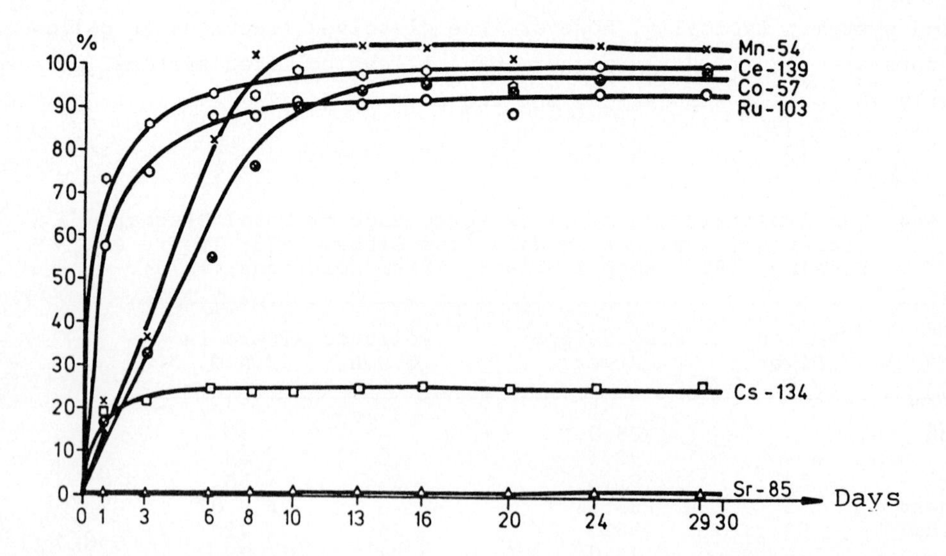

Figure 4-3 Sorption Kinetics of Different Radionuclides on Suspended Matter from Weser River (Schoer & Förstner, 1985)

Non-Linear Sorption, Irreversibility

Investigations on the reversibility of adsorption equilibria of heavy metals by Lion *et al.* (1982) indicate that their behaviour is strongly different in natural sediments compared to the experience on well-characterized solid surfaces. It is suggested that reduced reversibility for metals such as lead and copper is mainly the effect of **hydrolysis** and **specific adsorption** (Förstner, 1986).

With respect to the adsorption reaction between dissolved organic chemicals and naturally occurring sediments and suspended particles it has been shown in several laboratory experiments that measured desorption does not conform to the adsorption isotherm, and these **non-singular** or **hysteric isotherms** have been ascribed to a number of possible experiment artifacts. With their equilibration studies on sediment-adsorbed hexachlorobiphenyl (HCBP) DiToro & Horzempa (1982) were among the first to distinguish between reversible and strongly bound or resistant fractions of the same isomer. In the case of HCBP experimental evidence suggests that under certain chemical conditions the binding to sediments during consecutive desorption may be a **curvilinear isotherm** that may or may not ultimately demonstrate complete desorbability; it is also quite possible that the actual desorption process may involve binding to sites of a gradation of energies rather than the two arbitrarily defined fractions (reversible and resistant).

Chlorinated phenols, hydrophobic weak acids (category "b" in Chapter 2.3) exhibiting octanol/water partition coefficients between 10^2 and 10^5, constitute a class of compounds which is of growing concern. At relative low overall sorption rates pH is a dominant factor, and non-linear sorption isotherms were obtained which were interpreted to be the result of the superposition of several different sorption processes (Zierath et al., 1980). It has been stressed by Isaacson & Frink (1984) that partial irreversibility of "sorption" on sediments of phenol compounds is affected by the **penetrability** of sediment-associated organic matter. On the one hand, an increase in the degree of - pH-dependent - **dissociation** of the sorptive species would tend to decrease its hydrophobicity and increase its effective size by increasing the extent of its **hydration**. On the other hand, an increase in pH would also tend to increase the dissociation of weakly acidic functional groups on the sediment organic matter, which in turn could cause some "opening out" of these substrates and an increased availability of access to some sorption sites.

Bioconcentration Effects

Methods for estimating degree of sorption (distribution ratio) for organic compounds on sorbing substrate work surprisingly well in many cases, particularly at higher concentrations of organic solid matter. However, further work is needed to establish quantitative data for partition coefficients of non-polar organic compounds with **organic-poor sorbents** (Farrington & Westall, 1986). This mainly refers to the interactions of organic contaminants in groundwater aquifers (Schwarzenbach *et al.*, 1983).

Another gap is obvious with respect to the **biogenic activity** which mediates the interactions between aqueous and solid phases of organic chemicals. An example of the possible implications has been given by Oliver & Charlton (1984) for the behaviour of 1,4-dichlorobenzene in the Niagara River area. From the field data of DCB-concentrations in sediments (see Chapter 2.4.) and the respective values for the water phase, which were 19 ng/L of 1,4-DCB in 6 samples from spring and autumn 1980 (Oliver & Nicol, 1982), the apparent distribution coefficient of approximately 10^4 calculated from these data would reflect - with assumed 10% organic carbon in the Niagara River suspended sediment - a log K_{OC} (real K_D) of approximately 5.0 (Table 4-5). From particulates settling in Lake Ontario a log K_{OC}-value for 1,4-DCB of 4.57 was determined (Oliver & Charlton, 1984). These values are two orders of magnitude higher than the data from laboratory experiments and fron calculations using the relationship with the octanol/water partition coefficient. It seems that a major part of the particulates in both river suspension and in the sediment trap of this area is biogenic, and it can be concluded that for the assessment of the natural partition of organic chemicals the effect of bioconcentration has to be considered.

Table 4-5 Comparison of Solid/Solution Partition of 1,4-Dichlorobenzene in Natural Samples, from Laboratory Experiments, and from Calculations with Octanol/Water Partition Coefficients

Suspended particles/water in Niagara River	$\log K_{OC} = 5.0$
Settling particulates/water in Lake Ontario	$\log K_{OC} = 4.57$
Calculated from octanol/water coefficient	$\log K_{OC} = 2.99$
Determined from laboratory experiments	$\log K_{OC} = 2.54$

4.3 Climatic, Geochemical and Seasonal Variabitities of K_D-Factors

Climatic Variability

From the work of Martin and Meybeck (1979) it is indicated that the most variable elements in a geographical sense are those either carried mostly in solution (Ca, Na, Mg, As, Sr, Li), or significantly enriched in the suspended matter (Zn, Pb, Mo, Cu); the latter group partly reflects anthopogenic influences. The most constant elements are rare earths, Si, Ti, Al, Fe, Sc, and U. Metal contents of river solids depend nearly exclusively on allochthonous influences; after the composition of source rocks the most decisive factor is grain size. In tropical rivers the contents of Al, Fe, and Ti in the suspended matter are higher than in those from temperate and actic climate, where the contents of Ca are higher; for the former type of rivers the particulates mainly originate from soil material enriched in insoluble rather than the more soluble elements whereas for the latter type of rivers the suspended load is mainly derived from rock debris or poorly weathered particles.

Generally, variations of trace metal discharge are mainly caused by the changes in **suspended matter concentrations**. The relative importance of the particulate forms in the decreasing sequence Al > Fe > Mn > Co > Zn > Ni > Cu > Cd has been found by Yeats & Bewers (1982) from St. Lawrence River in Canada. According to these authors the lower proportions of the metals in particulate phases from St. Lawrence compared ato the data from the Amazon (Gibbs, 1977) or the Mississippi (Trefry & Presley, 1976) rivers are a direct reflection of the contrasting levels of suspended matter in the St. Lawrence (10 mg/l), Amazon (65 mg/l) and Mississippi (580 mg/l).

Geochemical Variability

With respect to the geochemical variability of trace elements, a study on recent sediments from 74 lakes in different climatic regions (Förstner, 1977) indicates relative low variation coefficients for zinc and copper, followed by lead, mercury, cadmium and cobalt, whereas manganese, chromium and nickel are more variable in these materials. For the latter two elements, **lithogenic effects** ("granitoid" or "gabbroid" source rocks) are dominant, while the effect of weathering, including diagenetic mobilisation, typically influences the concentrations of manganese in the aquatic particulate matter; as a result of leaching processes, concentrations of manganese, iron, cobalt and zinc are reduced (examples from the Amazon basin). Civilisational influences are reflected by elevated contents of lead, cadmium, mercury and zinc. With respect to dissolved trace metals typical elements indicating recent volcanic activity are mercury (Weissberg and Zobel, 1973) and arsenic. For example, a comparison of the arsenic levels in the upper and lower Rhine indicates the latter to be "polluted", but the lower Rhine has still less arsenic than some "unpolluted" rivers in the Andes mountains (Andreae, 1978). Gorham & Swaine (1965) have shown that oxidized and reduced muds from the freshwater Lake Windermere differ little in elemental composition. The reduced muds muds are enriched to a small extent in Co, Ni, Pb, and Sn, but not in Ag, Cu, Mo or Zn.

Seasonal Variabilities

Metal concentrations in rivers, particularly in solid matter, are strongly influenced by the runoff. From studies on two Cornish estuaries it is suggested by Boyden *et al.* (1978) that higher trace element concentrations in high and low water surface and bottom samples in winter compared to summer probably reflect increased weathering and transport in the catchments during this season. Trace metal concentrations in the Susquehanna River well correlated with the amount of solids discharged (Carpenter *et al.*, 1975); when data are calculated for weight concentrations of metals in the solid fraction, it is found that all metals generally peak during December and January and secondary peaks occur for Co, Cr, Ni, Cu and Mn in July. Troup and Bricker (1975) suggest that this effect is due to decaying **organic matter** which is abundant in the Susquehanna River during these two periods. Studies performed by Grimshaw *et al.* (1976) on the River Ystwyth in mid-Wales, where strong metal pollution from past mining operations is still obvious, indicate that metal concentrations in solution are highest during low water flow periods, suggesting a dilution effect; for brief periods during initial stages of storm runoff, there is a very significant increase of the metal concentrations in solution, apperently due to a **flushing effect**.

4.4. Case Study: Rhine River

The Rhine River is probably the most important river in Europe, hydrologically as well as economically. It is 1,236 km long, and it drains an area about 107,000 km^2. The average flow rate increases from 1,026 m^3 per second at Basel to 2,258 m^3 per second at Emmerich, the Dutch/-German border. Peak flow occurs after the snow melts in the Alps with a flow rate close to 5,500 m^3 per second. The number of people living in the catchment of the Rhine River is about 32 million. About 20 million are receiving drinking water directly or indirectly from the river. Intensively industrialized areas are found along the river as well as along its tributaries. Large chemical factories are located at Basel (e.g. Ciby-Geigy), at Ludwigshafen (BASF), near Frankfurt (Hoechst), and at Leverkusen (Bayer). Generally the riverwater pollution escales downstreams due to the gradually increasing density of the population and of industry (Figure 4-4).

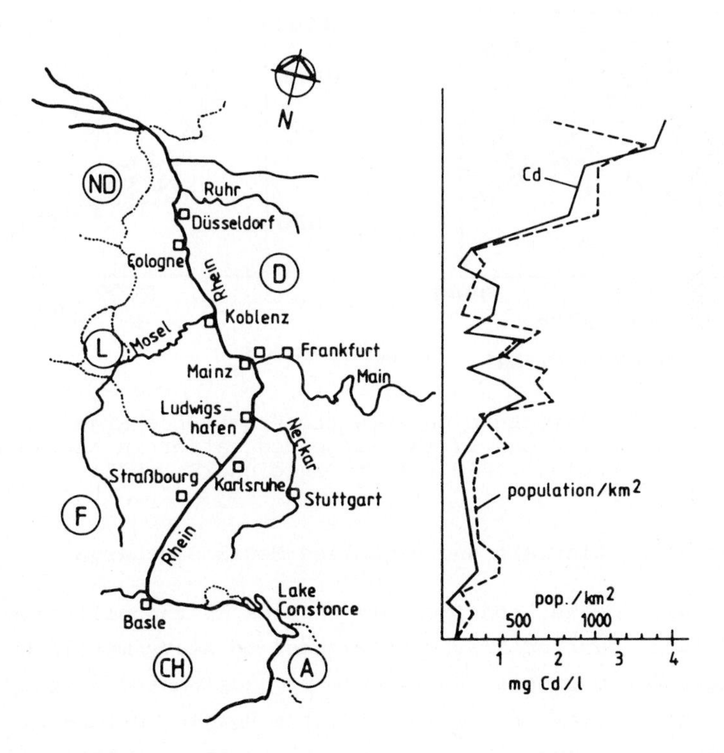

Figure 4-4 Population Density and Cadmium Concentration in the Catchment Area of the Rhine River (after Heinrichs, 1975)

Effect of Water Discharge

The major part of the heavy metal load in the lower Rhine River originates from anthropogenic sources. Mainly in the Netherlands, both dissolved and particulate metals cause tremendous problems, and the implications for dredging activities in Rotterdam Harbor will be presented in Chapter 7.4. Here, examples for the effect of water discharge on metal concentrations in particulate matter are shown in Figure 4-5 (after Schleichert, 1975). As indicated by chromium concentrations, there is a clear decrease of metal contents with increasing water flow. However, even the moderate cadmium contents in the winter period exceed the maximum values for the spring/summer, despite the much higher water discharge. Such a development is due to the suspended sediments, rich in cadmium, being held back in periods of poor flow, for example, mainly during the summer, in the lock-regulated Neckar and Main rivers; in autumn/winter the tendency is reversed, the material is carried by high water flows into the Rhine River in increased quantities.

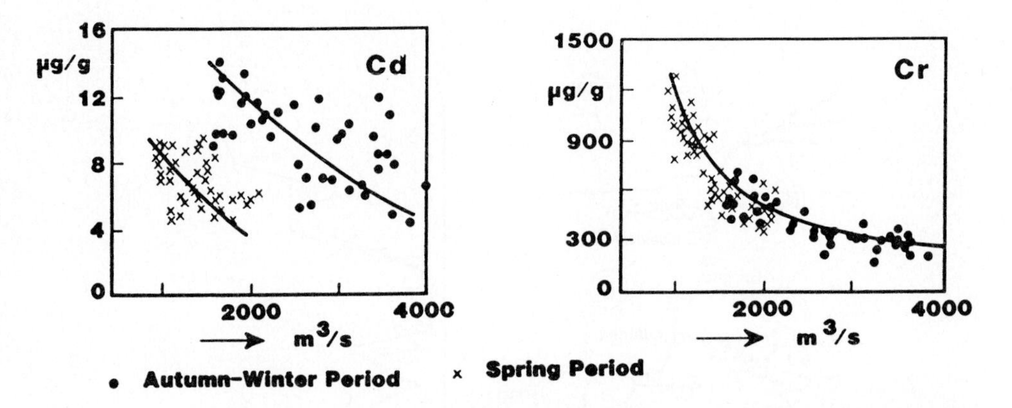

Figure 4-5 The Relationship between the Water Discharge and the Metal Concentrations in the Suspended Matter in the Rhine River near Koblenz, FRG (after Schleichert, 1975)

Development of Particulate and Dissolved Metal Discharge

During the last decade, considerable changes in the pollution status of the Rhine River, particularly with respect to heavy metals have taken place. Figure 4-6 presents the example of cadmium and mercury discharges in the Rhine River at the German/Dutch border. Discharges of cadmium have decreased from 250 tons in 1971 to approximately 50 tons in 1983. Mercury is even more effectively reduced from 100 tons to approximately 10 tons during this period (Malle, 1985).

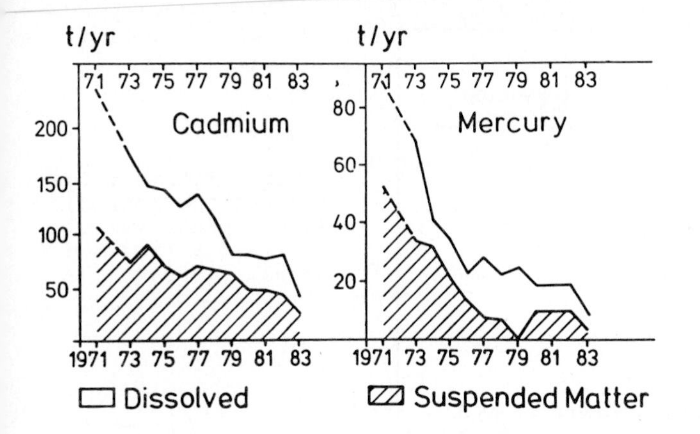

Figure 4-6

Changes of Metal Load (Particulate and Dissolved) in the Rhine River at the Dutch-German Border from 1971 to 1983 (after Malle, 1985)

This development is due to various factors, such as high water flow in "wet years" and the effects of the economic crisis, particularly at the end of the 1970's. However, a significant portion of the reduction should be affected by improvement of wastewater treatment and by the partial replacement of metals in critical applications. It is indicated from these data, that the major decrease for cadmium occurred in the dissolved phases, whereas - until 1979 - the reduction of mercury concentrations mainly took place in the solid phases. This is an indication that equilibria between solid and aqueous phases have not completely established, and clearly shows the difficulties involved in the modeling such processes (see review by Honeyman & Santschi, 1988).

Pesticide Release from "Sandoz"-Spill

The November 1, 1986, fire at a Sandoz Ltd. storehouse at Schweizerhalle, an industrial area near Basel, Switzerland, resulted in chemical contamination of the Rhine River and caused massive kills of benthic organisms and fish (Capel *et al.*, 1988). The majority of the more than 1300 tons of stored chemicals was destroyed in the fire, but large quantities were introduced into the Rhine River through runoff of the fire-fighting water. The chemicals with the highest measured concentrations in the river (at station Village-Neuf, 14 km downstream from the input point, at 15:15 on Nov. 1, 1986) were from the group of organophosphate pesticides, namely disulfoton (600 µg/L) and thiometon (500 µg/L). Model estimates have been made on the sediment pesticide concentrations and Figure 4-7 shows the dynamic sediment response to the Sandoz spill plume in the vicinity of the fish kill (Mossman *et al.*, 1988)

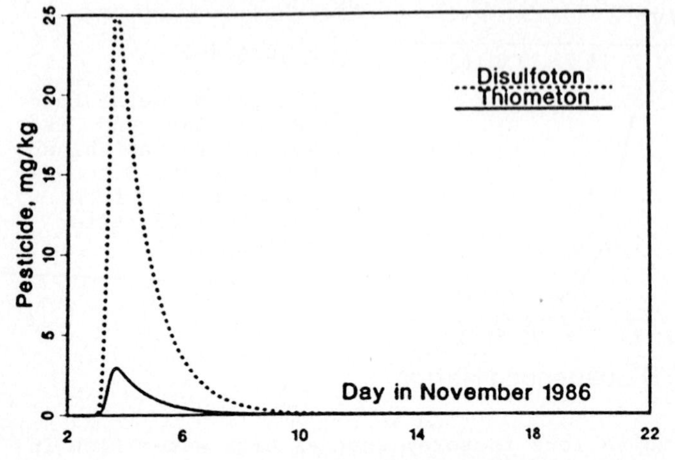

Figure 4-7

Dynamic Sediment Concentrations (mg/kg) 95 km Downstream of Spill Site (Mossman *et al.*, 1988)

Calculations by Hellmann (1987) using a solid/aqueous distribution coefficient for disulfoton of 3100 - as determined from model experiments with clay minerals - result in a sediment concentration of 1800 mg/kg in equilibrium with 600 µg/L (Figure 4-8); field measurements, however, on fine-grained sediment gave only 41 mg/kg. At Mainz, where the aqueous concentration of disulfoton was diluted to 20 µg/L, on would calculate respective 60 mg/kg in the sediment phase; de facto, field concentrations were less than 0.1 mg/kg in the sediment (in good agreement to the model in Figure 4-8). These data again demonstrate the problematic way of calculating equilibrium concentrations from laboratory data.

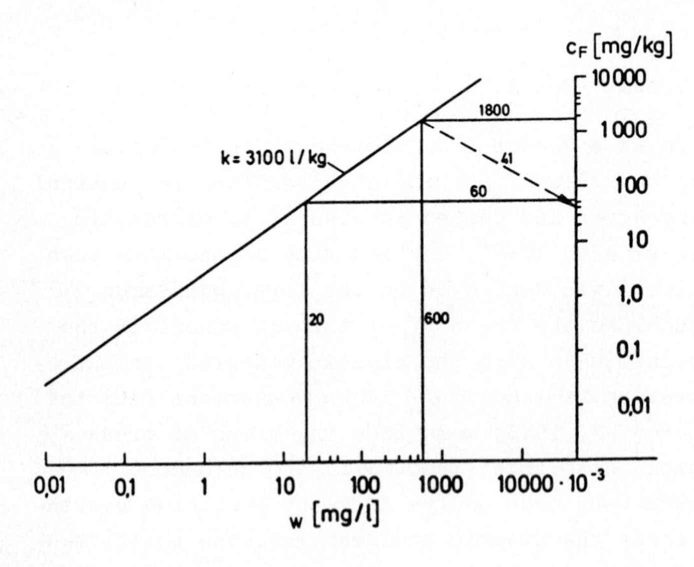

Figure 4-8

Diagram for Calculations of Equilibrium Concentrations between Dissolved and Sediment Fractions of Disulfoton following the "Sandoz-Spill" in the Rhine River near Basel (Hellmann, 1988)

References to 4.1 "Particles in Water"

Allan, R.J. (1986) *The Role of Particulate Matter in the Fate of Contaminants in Aquatic Ecosystems*. National Water Research Institute, Scientific Series No. 142, 128 p. Burlington: Canada Centre for Inland Waters

Baker, J.E., Chapel, P.D. & Eisenreich, S.J. (1986) Influence of colloids on sediment-water partition coefficients of polychlorophenyl congeners in natural waters. *Environ. Sci. Technol.* 20, 1136-1143.

Berner, R.A. (1981) A new geochemical classification of sedimentary environments. *J. Sediment. Petrol.* 51, 359-365.

Brownawell, B.J. & Farrington, J.W. (1986) Biogeochemistry of PCB's in interstitial waters of a coastal marine sediment. *Geochim. Cosmochim. Acta* 50, 157-169.

Carter, C.W. & Suffett, I.H. (1982) Binding of DDT to dissolved humic materials. *Environ. Sci. Technol.* 16, 735-740.

Chiou, C.T. *et al.* (1986) Water solubility enhancements of some organic pollutants and pesticides by dissolved humic and fulvic acids. *Environ. Sci. Technol.* 20, 502-508.

Davis, J.A. & Gloor, R. (1981) Adsorption of dissolved organics in lake water environments by aluminium oxide: Effect of molecular weight. *Environ. Sci. Technol.* 15, 1223-1227.

Frimmel, F.H. & Christman, R.F. (Eds.)(1988) *Humic Substances and Their Role in the Environment*. Dahlem-Konferenzen, Life Sci. Res. Rept. No. 41, 271 p. Chichester: Wiley.

Goldberg, E.D. (1976) *The Health of the Oceans*. 172 p. Paris: Unesco Press.

Gross, M.G. (1972) Geologic aspects of wasste solids and marine waste deposits, New York metropolitan region. *Geol. Soc. Amer. Bull.* 83, 3163-3176.

Hart, B.T. (1982) Uptake of trace metals by sediments and suspended particulates. In *Sediment/Freshwater Interactions*, ed. P.G. Sly, pp. 299-313. The Hague: Dr. W. Junk Publ.

Hassett, J.P. & Milicic, E. (1985) Determination of equilibrium and rate constants for binding of a polychlorinated biphenyl congener by dissolved humic substances. *Environ. Sci. Technol.* 19, 638-643

Jenne, E.A. (1977) Trace element sorption by sediments and soils - sites and processes. In *Symposium on Molybdenum*, eds. W. Chappell & K. Petersen, Vol. 2, pp. 425-553. New York: Marcel Dekker

Jones, B.F. & Bowser, C.J. (1978) The mineralogy and related chemistry of lake sediments. In *Lakes - Chemistry, Geology, Physics*, ed. A. Lerman, pp. 179-235. New York: Springer-Verlag.

Kranck, K. (1980) Sedimentation processes in the sea. In *The Handbook of Environmental Chemistry*, ed. O. Hutzinger, Vol. 2A, pp. 61-75. Berlin: Springer-Verlag.

McCarthy, J.F. & Jimenez, B.D. (1985) Interactions between polycyclic aromatic hydrocarbons and dissolved humic material binding and dissociation. *Environ. Sci. Technol.* 19, 1072-1076.

Müller, G. & Förstner, U. (1968) General relationship between suspended sediment concentration and water discharge in the Alpenrhein and some other rivers. *Nature* 217, 244-245.

Pierce, R.H., Olney, C.E. & Felbeck, G.T. jr. (1974) *pp'*-DDT adsorption to suspended particulate matter in sea water. *Geochim. Cosmochim. Acta* 38, 1061-1073.

Poirrier, M.A., Bordelon, B.R. & Laseter, J.L. (1972) Adsorption and concentration of dissolved carbon-14 DDT by coloring colloids in surface waters. *Environ. Sci. Technol.* 6, 1033-1035.

Postma, H. (1967) Sediment transport and sedimentation in the marine environment. In *Estuaries*, ed. G.H. Lauff, *AAAS-Publ.* 83, 158-179.

Raudkivi, A.J. (1976) *Loose Boundary Hydraulics*. Oxford: Pergamon

Salomons, W. (1987) *Sediment Pollution in the EEC*. Internal Report T 244, 136 p. Haren: Delft Hydraulics Laboratory/Institute for Soil Fertility.

Salomons, W. & Baccini, P. (1986) Chemical species and metal transport in lakes. In *The Importance of Chemical "Speciation" in Environmental Processes.*, eds. M. Bernhard *et al.*, pp. 193-216. Berlin: Springer-Verlag.

Shea, D. (1988) Developing national sediment quality criteria. *Environ. Sci. Technol.* 22, 1256-1260.

Sly, P.G. (1978) Sedimentary processes in lakes. In *Lakes - Chemistry, Geology, Physics*, ed. A. Lerman, pp. 65-89. New York: Springer-Verlag.

Tipping, E. (1981) The adsorption of aquatic humic substances by iron hydroxides. *Geochim. Cosmochim. Acta* 45, 191-199.

Weber, J.H. (1988) Binding and transport of metals by humic materials. In *Humic Substances and Their Role in the Environment*, eds. F.H. Frimmel & R.F. Christman, Dahlem-Konferenzen, Life Sci. Res. Rep. No. 41, 271 p. Chichester: Wiley.

Wolfe, R.G. (1964) The dearth of certain sizes of materials in sediments. *J. Sediment. Petrol.* 34, 320-327.

References to 4.2 "Factors Influencing Distribution Coefficients"

Allan, R.J. (1986) *The Role of Particulate Matter in the Fate of Contaminants in Aquatic Ecosystems.* National Water Research Institute, Scientific Series No. 142, 128 p. Burlington: Canada Centre for Inland Waters

Anderson, M.A. (1981) Kinetics and equilibrium control of interfacial reactions involving inorganic ionic solutes and hydrous oxide solids. In *Environmental Speciation and Monitoring Needs for Trace Metal-Containing Substances from Energy-Related Processes*, eds. F.E. Brinckman & R.H. Fish, pp. 146-162. Washington D.C.: U.S. Dept. of Commerce.

Benjamin, M.M. & Leckie, J.O. (1981) Multiple-site adsorption of Cd, Cu, Zn, and Pb on amorphous iron oxyhydroxide. *J. Colloid Interface Sci.* 79, 209-211.

Benjamin, M.M., Hayes, K.L. & Leckie, J.O. (1982) Removal of toxic metals from power-generated waste streams by adsorption and coprecipitation. *J. Water Pollut. Control Fed.* 54, 1472-1481.

Bowen, H.J.M. (1977) Residence times of heavy metals in the environment. Proc. Intern. Conf. *Heavy Metals in the Environment, Toronto,* 1975. Vol. I, pp. 1-9. Toronto: University of Toronto

Calmano, W. (1979) *Untersuchungen über das Verhalten von Spurenelementen an Rhein- und Mainschwebstoffen mit Hilfe radioanalytischer Methoden.* Doctoral Dissertation TH Darmstadt.

Chen, Y.R., Butler, J.N. & Stumm, W. (1973) Kinetic study of phosphate reaction with aluminium oxide and kaolinite. *Environ. Sci. Technol.* 7, 327-332.

Connor, D.J. & Connolly, J.P. (1980) The effect of concentration of adsorbing solids on the partition coefficient. *Water Res.* 14, 1517-1523.

DiToro, D.M. & Horzempa, L.M. (1982) Reversible and resistant components of PCB adsorption-desorption: Isotherms. *Environ. Sci. Technol.* 16, 595-602

Duursma, E.K. (1984) Problems of sediment sampling and conservation for radionuclide accumulation studies. In *Sediments and Pollution in Waterways.* IAEA-TecDoc-302, pp. 127-135. Vienna: International Atomic Energy Agency.

Farrington, J.W. & Westall, J. (1986) Organic chemical pollutants in the oceans and groundwater: A review of fundamental chemical properties and biogeochemistry. In *Role of the Ocean as a Waste Disposal Option*, ed. G. Kullenberg, pp. 361-425. Dordrecht: D. Reidel Publ.

Gibbs, R.J. (1977) Transport phases of transition metals in the Amazon and Yukon Rivers. *Geol. Soc. Amer. Bull.* 88, 829-843.

Förstner, U. (1984) Metal pollution of terrestrial waters. In *Changing Metal Cycles and Human Health*, ed. J.O. Nriagu, Dahlem-Konferenzen, pp. 71-94. Berlin: Springer-Verlag

Förstner, U. (1986) Solid/solution relations of contaminants in surface waters. In Proc. Euromech 192 "*Transport of Suspended Solids in Open Channels*", ed. W. Bechteler, pp. 209-220. Rotterdam: A.A. Balkema Publ.

Heinrichs, H. (1975) *Die Untersuchuing von Gesteinen und Gewässern auf Cd, Sb, Hg, Tl, Pb und Bi mit der flammenlosen Atomabsorptions-Spektralphotometrie*. Doctoral Thesis, Univ. Göttingen, 82 pp.

Isaacson, P.J. & Frink, C.R. (1984) Nonreversible sorption of phenolic compounds by sediment fractions: The role of sediment organic matter. *Environ. Sci. Technol.* 18, 43-48.

Kopp, J.F. & Kroner, R.C. (19068) *Trace Metals in Waters of the United States.* Fed. Water Pollut. Control Administration, Div. Pollut. Surveillance.

Lion, L.W., Altman, R.S. & Leckie, J.O. (1982) Trace-metal adsorption characteristics of estuarine particulate matter: Evaluation of contribution of Fe/Mn oxide and organic surface coatings. *Environ. Sci. Technol.* 16, 660-666.

Oliver, B.G. & Charlton, M.N. (1984) Chlorinated organic contaminants on settling particulates in the Niagara River vicinity of Lake Ontario. *Environ. Sci. Technol.* 18, 903-908

Oliver, B.G. & Nicol, K.D. (1982) Chlorobenzenes in sediments, water and selected fish from Lakes Superior, Huron, Erie and Ontario. *Environ. Sci. Technol.* 16, 532-536.

Schoer, J. & Förstner, U. (1985) Chemical forms of artificial radionuclides in fluviatile, estuarine and marine sediments compared with their stable counterparts. In Proc. Seminar on the *Behaviour of Radionuclides in Estuaries*, ed. E.K. Duursma, pp. 27-54. Luxembourg: Commission of the European Communities.

Schwarzenbach, R.P. *et al.* (19083) Behavior of organic compounds during infiltration of river water to groundwater: Field studies. *Environ. Sci. Technol.* 17, 472-479.

Trefry, J.H. & Presley, B.J. (1976) Heavy metal transport from the Mississippi River to the Gulf of Mexico. In *Marine Pollution Transfer*, ed. H.L. Windom & R.A. Duce, pp. 39-76. Lexington: D.C. Heath Publ.

Voice, T.C., Rice, C.P. & Weber, W.J. jr. (1983) Effects of solids concentrations on the sorptive partitioning of hydrophobic pollutants in aquatic systems. *Environ. Sci. Technol.* 17, 513-518.

Zierath, D. *et al.* (1980) Sorption of benzidine by sediments and soils. *Soil Sci.* 129, 277-286.

References to 4.3 "Variabilities"

Andreae, M.O. (1978) Disstribution and speciation of arsenic in natural waters and some marine algae. *Deep-Sea Res.* 25, 391-398.

Boyden, C.R., Aston, S.R. & Thornton, I. (1979) Tidal and seasonal variations of trace elements in two Cornish estuaries. *Estuar. Coastal Mar. Sci.* 9, 303-317.

Carpenter, J.H., Bradford, W.L. & Grant, V. (1975) Processes affecting the composition of estuarine water (H_2CO_3, Fe, Mn, Zn, Cu, Ni, Cr, Co, and Cd). In *Estuarine Research*, ed. L.E. Cronin, Vol. 1, pp. 137-152. London: Academic Press

Förstner, U. (1977) Metal concentrations in recent lacustrine sediments. *Arch. Hydrobiol.* 80, 172-191.

Gibbs, R.J. (1977) Transport phases of transition metals in the Amazon and Yukon Rivers. *Geol. Soc. Amer. Bull.* 88, 829-843.

Gorham, E. & Swaine, D.J. (1965) The influence of oxidizing and reducing conditions upon the distribution of some elements in lake sediments. *Limnol. Oceanogr.* 10, 268-279.

Grimshaw, D.L., Lewin, J. & Fuge, R. (1976) Seasonal and shortterm variations in the concentration and supply of dissolved zinc to polluted aquatic environments. *Environ. Pollut.* 11, 1-7.

Martin, J.M. & Meybeck, M. (1979) Elemental mass balance of material carried by major world rivers. *Mar. Chem.* 7, 173-206.

Trefry, J.H. & Presley, B.J. (1976) Heavy metal transport from the Mississippi River to the Gulf of Mexico. In *Marine Pollution Transfer*, ed. H.L. Windom & R.A. Duce, pp. 39-76. Lexington: D.C. Heath Publ.

Troup, B.N. & Bricker, O.P. (1975) Processes affecting the transport of materials from continents to the ocean. In: *Marine Chemistry in the Coastal Environment*, ed. T.M. Church. *ACS Symp. Ser.* 18, 133-151. Washington, D.C.: American Chemical Society

Weissberg, B.G. & Zobel, M.G. (1973) Geothermal mercury pollution in New Zealand. *Bull. Environ. Contam. Toxicol.* 9, 148-155.

Yeats, P.A. & Bewers, J.M. (1982) Discharge of metals from the St. Lawrence River. *Can. J. Earth Sci.* 19, 982-992

References to 4.4 "Case Study: Rhine River"

Capel, P.D. *et al.* (1988) Accidental input of pesticides into the Rhine River. *Environ. Sci. Technol.* 22, 992-997.

Heinrichs, H. (1975) *Die Untersuchuing von Gesteinen und Gewässern auf Cd, Sb, Hg, Tl, Pb und Bi mit der flammenlosen Atomabsorptions-Spektralphotometrie*. Doctoral Thesis, Univ. Göttingen, 82 pp.

Hellmann, H. (1987) Organische Spurenstoffe im Dreiphasensystem Wasser-Schwebstoff-Luft; eine Einführung. *Vom Wasser* 69, 11-22.

Honeyman, B.D. & Santschi, P.H. (1988) Metals in aquatic systems - predicting their scavenging residence times from laboratory data remains a challenge. *Environ. Sci. Technol.* 22, 862-871.

Malle, K.-G. (1985) Metallgehalt und Schwebstoffgehalt im Rhein. *Z. Wasser Abwasser Forsch.* 18, 207-209.

Mossman, D.J., Schnoor, J.L. & Stumm, W. (1988) Predicting the effects of a pesticide release to the Rhine River. *J. Water Pollut. Control Fed.* 60, 1806-1812.

Schleichert, U. (1975) Schwermetallgehalte der Schwebstoffe des Rheins bei Koblenz im Jahresablauf. *Dtsch. Gewässerkundl. Mitt.* 19, 150-157.

Themes for Further Consideration

"Physico-Chemical **Forms of Trace Elements** in Solution, on Colloids and on Particulate Matter in Natural Waters, including Kinetic Effects"

"Interactions between Toxic **Organic Chemicals** and Sediment Particles"

"**Modeling the Transport** of Particle-Associated Contaminants in Rivers, Estuaries, Coastal Zones, Lakes, Reservoirs, Ponds, and Dumping Areas"

5. TRANSFER PROCESSES

5.1. In-Situ Processes

In Table 5-1, the major processes influencing the **cycling of contaminants** in aquatic systems are arranged according to the primary research discipline involved, and phase (dissolved or particuluate). There are interactions between chemistry and biology in the case of bioturbation, and between chemistry and physics for photodegradation.

Table 5-1 Processes Affecting Cycling of Pollutants in Aquatic Systems

	Aqueous Species	Particulate Phases
"Chemical"	Dissolution Desorption Complexation	Precipitation Adsorption Aggregation
	Species Transformations	
"Biological"	Decomposition Absorption, release Cell wall exchange	Food web transfer Filtering, digestion Pellet generation
	Bioturbation	
"Physical"	Advection Diffusion Photolysis	Resuspension Settling Burial

Biological activity is involved in physical cycling of particulate matter both in the water column and at the sediment/water interface. Organic exretions may produce fecal pellets and may enhance aggregation and thus faste settling of particles (Honjo, 1980). There are well-documented effects of reworking and resuspension of sediments by benthic organisms such as tubifid worms, but also by amphipods, shrimps, and clams. **Bioturbation** is a major post-sedimentation process (Petr, 1977), affecting the fate of particle-associated toxic metals and persistent organic chemicals, which are not primarily affected by volatilization, photolysis or bio- and photo-degradation (Allan, 1986).

Figure 5-1 shows the interactions of pollutants in various **subsystems**, including atmosphere, water column, active sediment, and deep sediment (Eadie *et al.*, 1983).

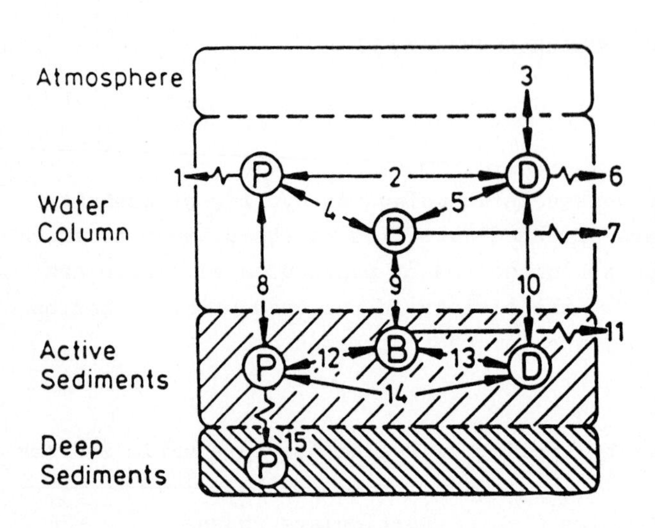

Figure 5-1

Compartments and Interactions of Toxic Chemicals (after Eadie *et al.*, 1983, and Allan, 1986).
P = Particular Phase;
D = Dissolved Phase;
B = Abbreviated Food Web
Processes: 1 & **6** = Photolysis; **2** & **14** = Sorption; **3** = Air-Water Exchange; **4** = Grazing and Fecal Pellet Generation; **5** = Filtering; **7** & **11** = Biological Decomposition; **8** = Settling and Resuspension; **9** = Food Web Dynamics; **10** = Affective and Diffuse Mixing; **12** & **13** = Benthos-Sediment-Interaction; **15** = Burial and Bioturbation

Whereas in Chapter 4 interactions between suspended particles and dissolved constituents have been described, Chapter 5 focusses on the processes taking place **within the sediment** or near the sediment/water interface. These processes depend on a large extent on the position of the **oxic-anoxic interface**. This interface (redoxcline) is within the sediments of well-mixed waterbodies and in the water column of some stratified lakes and basins. Transport processes and changes in the chemical environment determine the behaviour of contaminants (Salomons *et al.*, 1987).

Investigations in different Swiss lakes (Baccini, 1985) have led to a simple **two-box model** for the sediment boundary layer (Figure 5-2; Salomons & Baccini, 1986): The cucial zone is the oxic/anoxic interface where sharp redox gradients are observed. The transport of redox-sensitive species can be governed by these redox gradients. The classical examples are iron and manganese for which a simple conceptual model can well describe the qualitative aspects of transport phenomenolgy (Davison, 1985). Initial efforts for applying quantitative models have been undertaken both for the determination of the mobilizable fraction in the sediment (e.g., Baes & Sharp, 1983) and for the dynamics of pore water constituents (e.g., Schmitt & Sticher, 1986).

While these approaches, particularly with respect to trace metals, are gradually evolving, there is already much **experimental evidence**, that the newly formed iron(III) and manganese(IV) oxides offer new adsorbing surfaces. Upward diffusing trace metals encounter this freshly precipitated material and are adsorbed onto it, whereby the reflux of trace metals is reduced. This **iron barrier** (see Figure 5-2) moves upward with continued sedimentation (Hallberg, 1978). The chemical nature of these newly formed inorganic polymers is not yet fully understood, but it is clear that the presence of dissolved organic matter (concentration of dissolved organic carbon in interstitial waters of the sediment boundary layer is approx. 10^{-3} M) influences stongly the size and stability of these polymers.

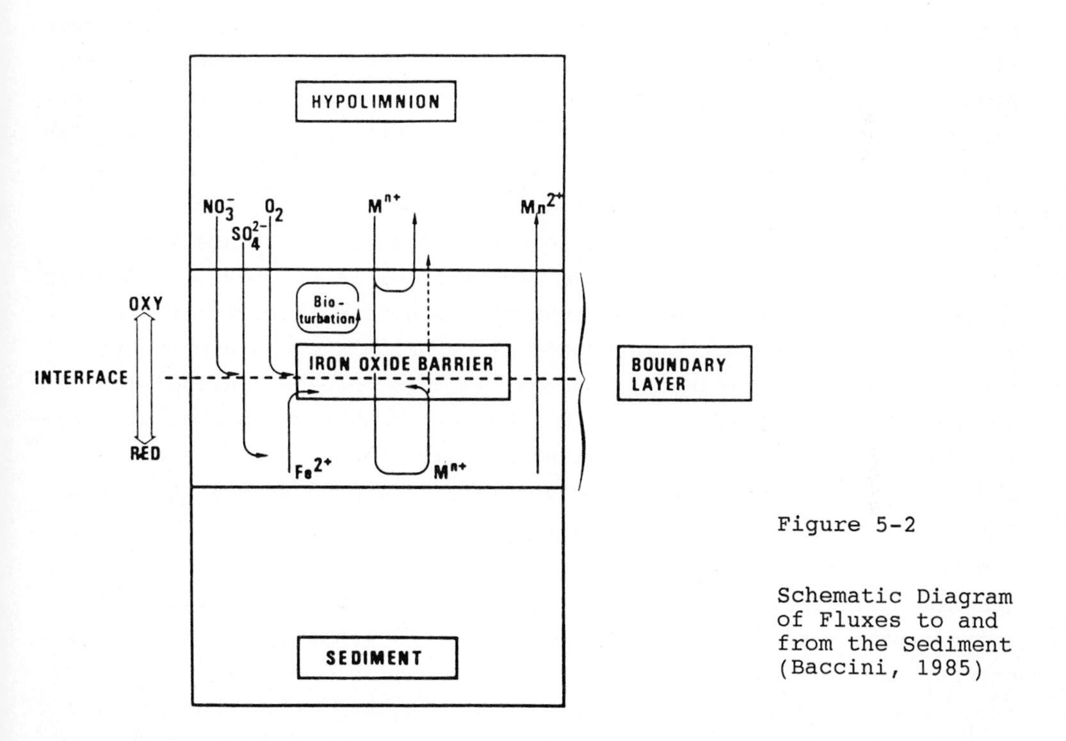

Figure 5-2

Schematic Diagram of Fluxes to and from the Sediment (Baccini, 1985)

The sediment layer above the redoxcline may also act as a **source** of metals for the overlying water. In the oxic surface layer a breakdown of the organic matter occurs resulting in a release of trace metals associated with solid organic phases. Furthermore, dissolved organic matter is produces, which may form complexes with the released trace metals or may actively desorb them from the sediment. Experiments by Maes & Cremers (1986) suggest that generation of ferric oxides during oxidation may lead to a displacement of trace metal from the humic acid sink.

It has been stressed by Salomons (1985) that from an **impact** point of view it import to know whether the concentrations in the pore waters are determined by **adsorption/desorption** processes or by **precipitation/-dissolution** processes. If the latter is the case the concentrations in the pore waters of pollutants are independent of the concentrations in the solid phase. The is strong direct (Luther *et al.*, 1980; Lee & Kittrick, 1984) and indirect (Lu & Chen, 1977) evidence that the concentrations of copper, cadmium and zinc in sulfidic pore waters are determined by precipitation-dissolution processes; the concentrations of arsenic and chromium in pore waters are probably controlled by adsorption-desorption processes, and mainly depend on the concentrations in the solid phase (Salomons, 1985).

Pore water data are playing a central role in the study and interpretation of processes affecting so-called "in-place" or **"in-situ" contaminants**. Significant enrichment of trace metals in pore waters has been found in (anoxic) sediment samples from Southern California basin, Saanich Inlet (Britsh Columbia) and Loch Fyne (Scotland), and has been explained by effects of **complexation** by organic substances (Brooks *et al.*, 1968; Presley *et al.*, 1972; Duchart *et al.*, 1973). Nissenbaum & Swaine (1976) found that with the exception of iron, nickel and cobalt (which mostly occur as sulfides under anaerobic conditions) the elements concentrated in the interstitial solution are those which are also enriched in sedimentary **humates**. In this respect Jonasson (1977) established a probable order of binding strength for a number of metal ions onto humic or fulvic acids: $Hg^{2+} > Cu^{2+} > Pb^{2+} > Zn^{2+} > Ni^{2+} > Co^{2+}$. The transport of dissolved metals in the pore waters is, therefore, strongly influenced by the vertical gradient of dissolved humic substances (Krom & Sholkovitz, 1978). On the other hand, it has been stressed by Salomons *et al.* (1987) that in pore waters **bisulfide ions** compete with the dissolved organic matter for complexation of the metals. Their calculations on the basis of available equilibrium partition values (Van den Berg & Dharmvanij, 1984) suggested that organic matter does not complex copper, zinc and cadmium to an appreciable extent in the presence of (bi)sulfide. From this controversy it becomes evident, that for a full understanding of the behaviour of trace metals in pore waters and their possible bioavailability more experimental speciation studies (Hart & Davies, 1977; Batley & Giles, 1980; Elderfield, 1981) and identification of solid phases are urgently needed.

5.2. Mobilization of Pollutants

"Mobilization", in a wide sense, comprises changes in the chemical environment which usually are affecting **lower rates of precipitation or adsorption** - compared to "natural" conditions - rather than active releases of contaminants from solid materials. With respect to particle-bound metals, their solubility, mobility and bioavailability can be increased by **five major factors** in terrestrial and aquatic environments:

Acidification. Acidity imposes problems in all aspects of metal mobilization in the environment: toxificity of drinking water, growth and reproduction of aquatic organisms, increased leaching of nutrients from the soil and the ensuing reduction of soil fertility, increased availability and toxicity of metals, and the undesirable acceleration of mercury methylation in sediments (Fagerström & Jernelöv, 1972). On a regional scale, acid precipitation is probably the prime factor affecting metal mobility in surface waters - by changing solid/dissolved equilibria in the atmospheric precipitations, washout effects on soils and rocks in the catchment area, enhancing groundwater mobility of metals and by active remobilization from aquatic sediments. In Swedish lakes, for example, a pronounced correlation was observed between dissolved metal levels and pH (Dickson, 1980). Acidic drainage from coal and ore mines has been recognized as a serious environmental pollution problem since long (review by Förstner, 1981). Transformations of sulfides and a shift to more acid conditions is particularly inhancing mobility of elements such as Mn, Fe, Zn, Pb, Cu, and Cd. A simple chemical mechanism could not explain the rapid production of acidic mine drainage by the oxidation of metal sulfides. According to Singer and Stumm (1970) ferric iron is the major oxidant of pyrite in the complex natural oxidation sequence and it is mainly *Thiobacillus ferrooxidans*, an iron-oxidizing acidophilic bacterium, which accelerates metal sulfide oxidations 10^6 times over the abiotic rate.

Salinity Increase. The effect of higher salinity seems to be particularly critical for resuspended cadmium-rich sediments in estuaries (Salomons & Förstner, 1984). As a result of biological or biochemical pumping, the tidal flats may act as a source of dissolved metals (Morris *et al.*, 1982). Release of trace metals from particulate matter has been reported from several estuaries (Scheldt, Gironde, Elbe/Weser, Savannah/Ogeechee), and has been explained by oxidation processes and by intensive breakdown of organic matter (both mediated by microorganisms), whereafter the released metals become complexed with chloride and/or ligands from the decomposing organic matter in the water. According to experimental data given by Salomons and Mook (1980) these effects can even be found in salt-polluted inland waters: at chloride contents of 200 mg/L - example: Lower Rhine river - the "normal" adsorption rate of cadmium would be reduced by approximately 20%; at 1.000 mg/L Cl^- - example: Weser River in Germany - this rate would be only half compared to the sorption of Cd under natural salt concentrations.

Complexing agents. Significant effects on the mobility of heavy metals can be expected by strong synthetic chelators, such as nitrilotriacetate (NTA), a substitute for polyphosphate in detergents, and ethylenediamintetracetate (EDTA), which is used as well for replacing phosphate, but also in metal-processing, galvanotechnology, and photo-industry. The extent of metal mobilization depend on the concentration of complexing agent, its pH-value, the mode of occurrence of heavy metals in the suspended sediment and on competition by other cations. Active remobilization seem to exhibit reliable results at NTA-concentrations above approx. 1-2 mg/L; such concentrations of NTA could rarely be expected in normal river waters, but may occur at even higher levels in sewage treatment plants. "Passive" effects of NTA (where the complexing agent may negatively influence the natural adsorption processes) are starting at lower values, at NTA concentrations of 200 to 500 μg NTA/l, and it has been found by Salomons (1983) that zinc adsorption is already significantly affected at NTA concentrations of 20 - 50 μg/L at pH-8 conditions.

Bio-Methylation. Organisms not only accumulate metals from the abiotic reservoirs but they are also able to interact with metals and modify processes affecting them, e.g. by pH increase, reduction of sulfate, redox conversions of inorganic forms, and release of extracellular material. Methylation of inorganic metal compounds is an important and well-known biogeochemical phenomenon in natural systems and has been shown to occur for a number of trace elements including Hg, As, and Sn (Craig, 1986). Mercury is the best studied example of an element that undergoes a complex cycle in the biosphere and for which there is evidence of the biochemical and molecular basis of the transformations (Summers & Silver, 1978). The product is mainly monomethylmercury under neutral and acidic conditions, and (volatile) dimethylmercury under basic conditions. Conversion of organic mercury to methyl mercury in anaerobic sediments is negatively correlated with salinity; as an explanation the theory is advanced that sulfide, derived from sea salt sulfate by microbial reduction, interferes with Hg^{2+}- methylation by forming highly insoluble HgS (Compeau & Bartha, 1983). Estimates for net methylation rates in sediments range from 15-40 ng/g per day to 137 ng/g per day, the latter in organic-rich salt marsh sediments (Windom, 1976).

Oxidation/reduction processes. Experimental data of Lu & Chen (1977) on the migration of trace metals between the interface of seawater and polluted surface sediment suggested that under reducing conditions the concentrations of trace metals are controled by sulfide complexes for Cd, Hg, and Pb. Under oxidizing conditions the controlling solid may change gradually from metallic sulfides to carbonates, oxyhydroxides, oxides, or silicates, thus changing the solubility of the associated trace metals.

Partition of a sediment sample from Hamburg Harbor, which was pretreated in different ways (EPA Standard Elutriate Test, 1:4 sediment/site water for 30 min.; freeze-dried sample; oven-drying at 60°C) clearly demonstrates the effect of oxidation in regulating the chemical form of cadmium and other trace metals (Figure 5-3; Kersten *et al.*, 1985): Compared to the original sample (**A**), which was extracted under an argon atmosphere, there is a typical change from oxidizable phases (mainly Cd-sulfide) to easily reducible forms upon application of the shaking/-bubbling test (**B**); during freeze-drying - which is commonly assumed to present a relative smooth mode of sample pretreatment - transformation to carbonatic and exchangeable forms takes place, and this effect is further enhanced during oven-drying at 60°C.

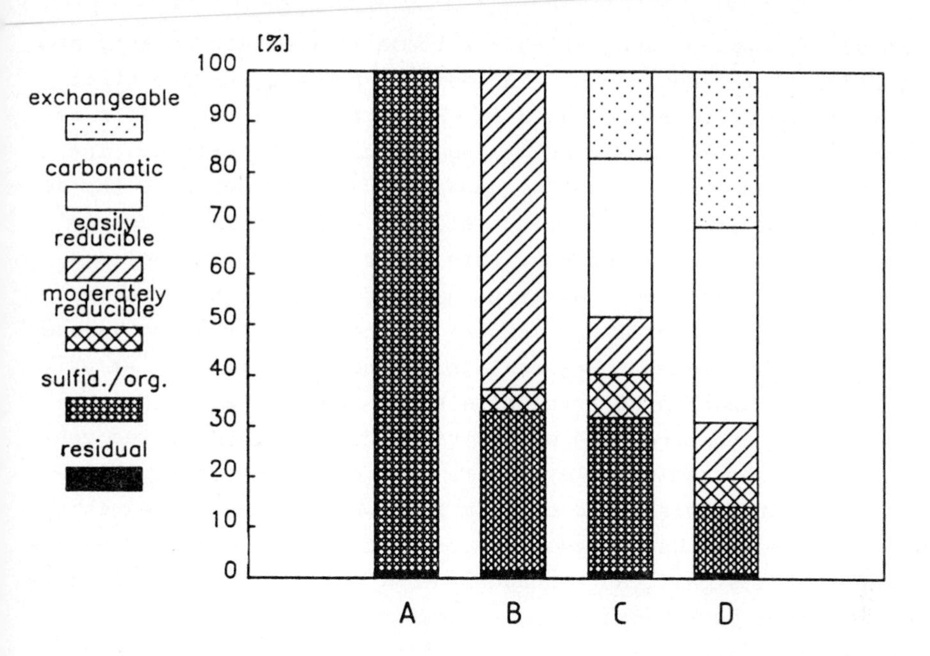

Figure 5-3 Partition of Cadmium in Anoxic Mud from Hamburg Harbor in Relation to the Pretreatment Procedures (Kersten *et al.*, 1985): (**A**) Control Extracted as Received Under Oxygen-Free Conditions; (**B**) After Treatment with Elutriate Test; (**C**) Freeze Dried; and (**D**) Oven-Dried (60°C)

Field evidence for changing **cadmium mobilities** was reported by Holmes *et al.* (1974) from Corpus Christi Bay Harbor; during the summer when the harbor water is stagnant cadmium is precipitated as CdS at the sediment/water interface; in the winter months, however, the increased flow of oxygen-rich water into the bay results in the release of some of the precipitated metal. Gendron *et al.* (1986) found evidence for different release mechanisms of cobalt and cadmium near the sediment-water interface in the St. Lawrence Estuary, in agreement with results from extraction procedures given in the case study on the Elbe River estuary (section 5.5). The hydroxylamine-extractable fraction of the (freeze-dried) sediment surface layer in the Lawrentian trough was enriched in cobalt but depleted in cadmium relative to the subsurface sediment. While in this layer the highest dissolved cadmium concentration was detected, the concentration of dissolved cobalt was relatively low. The profiles for cobalt ressemble those for manganese and iron with increased levels downwards, suggesting a release in the reduced zone and reprecipitation at the surface of the sediment profile. On the other hand, cadmium apperas to be released in the surface, probably as a result of the aerobic remobilization of organically bound cadmium. The authors found, in addition, that the upward flux of oxygen into the sediment and suggested that the cadmium released near the interface is recycled back into the water column. Pore water analyses in shallow-water sediments of Puget Sound estuary indicate higher concentrations of cadmium (as well as nickel and copper) in the upper oxidized layers of sediment covering reduced, H_2S-containing deposits (Emerson *et al.*, 1984). Such remobilization of trace metals has been explained by the removal of sulfide from the pore waters via biologically-mediated ventilation of the upper sediment layer with oxic overlying water, allowing the enrichment of dissolved cadmium that would otherwise exhibit very low concentrations due to the formation of insoluble sulfides. The authors suggested a significant enhancement of metal fluxes to the bottom waters by these mechanisms. It was evidenced by Hines *et al.* (1984) from **tracer experiments** that biological activity in surface sediments greatly enhances remobilization of metals by the input of oxidized water; these processes are more effective during spring and summer that during the winter months. From **enclosure experiments** in Narragansett Bay (Hunt & Smith, 1983) it has been estimated that be mechanisms such as oxidation of organic and sulfidic material, the anthropogenic proportion of cadmium in marine sediments is released to the water within approximately 3 years; for remobilization of copper and lead, approximately 40 and 400 years, respectively, is needed, according to these extrapolations.

Compared to the wide experience with mobilizing processes on trace metals knowledge on **desorption of organic contaminants** from solid substrates is still relatively poor. Regarding at first before-mentioned parameters controlling sorption/desorption processes of inorganic pollutants, it can be expected that **pH and ionic strength** of leachates should predominantly affect partitioning of **ionic organic compounds**. In fact, studies performed by Westall *et al.* (1985) indicate that for some environmentally important chlorinated phenols, i.e., for tetra- and pentachlorophenol, ionic strength and pH-values have to be considered in relation to octanol/water distribution.

5.3. Metal Transfer between Inorganic and Organic Substrates

With respect to the modeling metal partitioning between dissolved and particulate phases in a natural system, e.g. for estuarine sediments, the following reqirements have been listed by Luoma & Davis (1983):

- the determination of binding intensities and capacities for important sediment components,
- the determination of relative abundance of these components,
- the assessment of the effect of particle coatings and of multicomponent aggregation on binding capacity of each substrate,
- the consideration of the effect of major competitors (Ca^{2+}, Mg^{2+}, Na^{+}, Cl^{-}),
- the evaluation of kinetics of metal redistribution among sediment components.

It seems that **models** are still restricted because of various reasons: (i) adsorption characteristics are related not only to the system conditions (i.e., solid types, concentrations and adsorbing species), but also to changes in the net system **surface properties** resulting from particle/particle interactions such as coagulation; (ii) influences of **organic ligands** in the aqueous phase can rarely be predicted as yet; (iii) effects of **competition** between various sorption sites, and (iv) **reaction kinetics** of the individual constituents cannot be evaluated in a mixture of sedimentary components.

At present, **experimental studies** on the dissolved/solid interactions in such complex systems seem to be more promising. One approach is with a a six-chamber device, where the individual components are separated by membranes, which still permit phase interactions via solute transport of the elements (Calmano *et al.*, 1988); in this way, exchange reactions and biological uptake can be studied for individual phases under the influence of pH, redox, ionic strength, solid and solute concentration, and other parameters:

The **laboratory system** used in these studies was developed from the experience on sediment/algae interactions with a modified two-chambered device (Ahlf *et al.*, 1986). The system is made of a central chamber connected with 6 external chambers and separated by membranes of 0.45 µm pore diameter (Figure 5-4). The volume of the central chamber is litres and each of the external chambers contains 250 ml. Either solution or suspension can be inserted into the central chamber; in each external chamber the single solid components are kept in suspension by magnetic stirring. Redox, pH and other parameters may be controlled and adjusted in each chamber.

In an experimental series on the **effect of salinity**, i.e., disposal of anoxic dredged mud into sea water, quantities of model components were chosen in analogy to an average sediment composition: 0.5 g algal cell walls (=5%), 3 g bentonite (=30%), 0.2 g manganese oxide (=2%), 0.5 g goethite (=5%), and 5 g quartz powder (=50%). In the central chamber, 100 g of anoxic mud from Hamburg harbour was inserted; salts were added corresponding to the composition of sea water. After 3 weeks solid samples and filtered water samples were collected from each chamber and analysed by atomic absorption spectroscopy.

The effect of salinity on metal remobilization from contaminated sediments is different for the individual elements. While approximately 16% and 9% of **cadmium** and **zinc**, respectively, in the dredged mud from Hamburg harbour is released, for metals such as **copper** the factor salinity increase seems to be less important in the transfer both among sediment substrates and to aquatic biota. This is, however, not true as can be demonstrated from a mass balance for the element copper in Figure 5-5: It is indicated that only 1.3 % of the inventory of copper of the sludge sample is released when treating with seawater. Only one third stays in solution, equivalent to approx. 40 $\mu g l^{-1}$, and there is no significant difference to the conditions before salt addition. Two thirds of the released copper is readsorbed at different affinities to the model substrates. Quartz has just 3 ppm of copper and bentonite clay contains approx. 15 ppm, i.e., copper concentrations in these substrates are not significantly different from their natural contents. Slight enrichment of copper is observed in the iron hydroxide (approx. 80 ppm) and manganese oxide (100 ppm), whereas the cell walls - a minor component in the model sediment - has accumulated nearly 300 ppm of copper.

The dominant role of organic substrates in the binding of metals such as Cd and Cu is of particular relevance for the transfer of these elements into **biological systems**. It can be expected that even at relatively small percentages of organic substrates these materials are primarily involved in metabolic processes and thus may constitute the major carriers by which metals are transferred within the food chain.

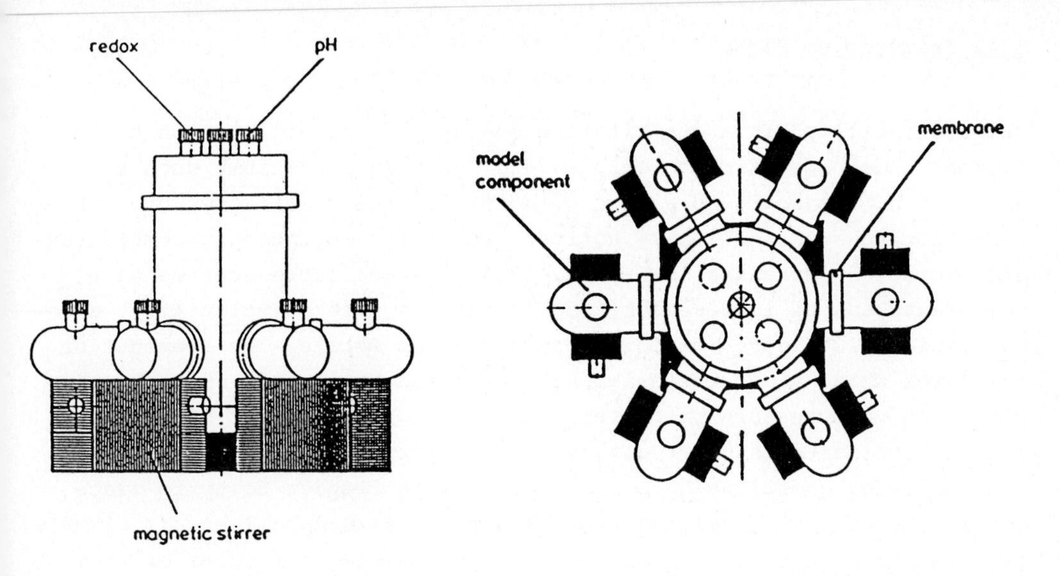

Figure 5-4 Schematic View of the Multi-Chamber Device (Calmano et al., 1988)

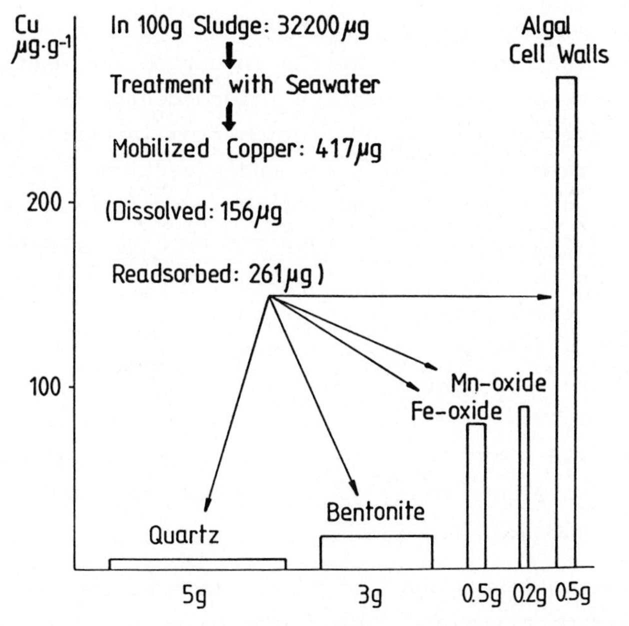

Figure 5-5 Transfer of Copper from Anoxic Harbor Mud into Different Model Substrates after Treatment with Artificial Seawater (after Data from Calmano et al., 1988)

5.4. Transfer to Biota

"Bioavailability" is a quantitative measure of the utilization of an element under specific conditions, and includes mechanisms such as absorption, transport to a site of metabolic/toxic activity, biotransformation to a metabolically active/toxic form, retention/accumulation, and excretion (McKenzie in Nriagu, 1984). Bioavailable species of a pollutants can be increased by small particle size or volatility, aqueous/lipid solubility, complexation, and - for metals - by elements or complexes which mimic essential nutrients and thus are handled by specific active transport processes. Assessment of pollutant bioavailability in sediments is difficult due the compexicity of the system, and the following questions have been raised with respect to toxic trace metals (Luoma, 1983; Salomons & Förstner, 1984; Campbell *et al.*, 1988): (i) Are pollutants been taken up from the ingested particles or from the pore waters? (ii) Is the uptake mediated by the food (e.g. bacteria) which accumulate metals from the sediment and/or the pore waters? (iii) Are the processes in the organism dominant or the binding of the metals to the sedimentary phases?

Toxic organic chemicals and trace metals in bottom sediments are accumulated by organisms by **direct sorption (bioconcentration)** - for example, by benthic invertebrates by adsorption from interstitial water to the body wall and respiratory surfaces, and ingestion of sediment and water through the gut - and **biomagnification**, i.e. concentration from consumption of lower trophic level organisms by higher trophic organics with a net increase in tissue concentrations (Munawar *et al.*, 1984). Bioaccumulation of **trace metals** by invertebrates, as expressed by the factor "tissue concentration/sediment concentration" (CF-values) generally has been found less than 1.0, except for mercury (Prosi, in Förstner & Wittmann, 1979, pp. 271-323), whereas for persistent **organic chemicals** such as PAH's and PCB's generally higher CF-values - from 2 to more than 10 - have been measured in benthic organisms (Reynoldson, 1987).

Particular interest has been devoted to the interactions of **PCBs** among benthic fauna, sediment and pore water, and based on field and laboratory tests, the following factors have been evidenced to affect its **bioaccumulation:** (i) toxic chemical structure (Goerke & Ernst, 1977); (ii) chemical concentration in the sediment (Fowler *et al.*, 1978); and (iii) sediment characteristics, especially organic content and particle size (Lynch & Johnson, 1982). PCB body burden in fish exposed to contaminated sediments results from the desorption of PCB to the water phase, the direct contact of fish with the sediment, the digestion of sediments, or from a combination of these processes (Califano *et al.*, 1982). Generally, food chain routes are more important than water uptake, and it has been suggested by Thomann & Connolly (1984) that food chain transfer accounts for more than 99% of the body burden of adult trout.

Both **biomagnification** and **detoxification** depend from many influences such as the ecological situation of an organism (e.g., choice of food, choice of macro- and microhabitat, and extent and periodicity of activity), the digestive physiology of the organism or parts of it, and effect of other chemical constituents (stimulation or competitive inhibition of uptake) both at the environmental interface of an organism or within tissues. In particular the differences among individuals in the formation of binding agents for metals may offer a partial explanation for the range of element concentrations in organs exposed to toxic doses of trace elements (Jenne & Luoma, 1977).

Whereas bioaccumulation of toxic organic substances appears to be based on relative simple mechanisms, mainly correlated to the **lipophility** of the compound and **organic carbon** concent in the sediment (Chapter 2.3) bioaccumulation of inorganic substances has to take into account many influencing factors, and is mainly complicated by competing ions. The different **forms of metals** in the growth medium and their effects on the uptake of metals in organisms are schematically summarized in Figure 5-6. It has been stressed by Baudo (1982) that the last three forms pertain to the "consumer" organisms alone, whereas the first three are equally shared with the "producer". The particulate form of the element can assume an important role in the uptake processes, at least for some kinds of organisms, such as filter feeders and deposit feeders. Conversely, the biota may interact in a number of ways with its surrounding medium, modifying the pseudo-equilibria among chemical species in the water/solid phases (Smies, 1983).

Estimation on the remobilization of metals under changing environmental conditions and on the **potential uptake by biota** are two major objectives of species differentiation on particle-bound trace metals. However, many authors have shown that with respect to bioavailability, as distinct from geochemical mobility, the present state of knowledge on solid matter speciation of metals is still somewhat unsatisfactory. The

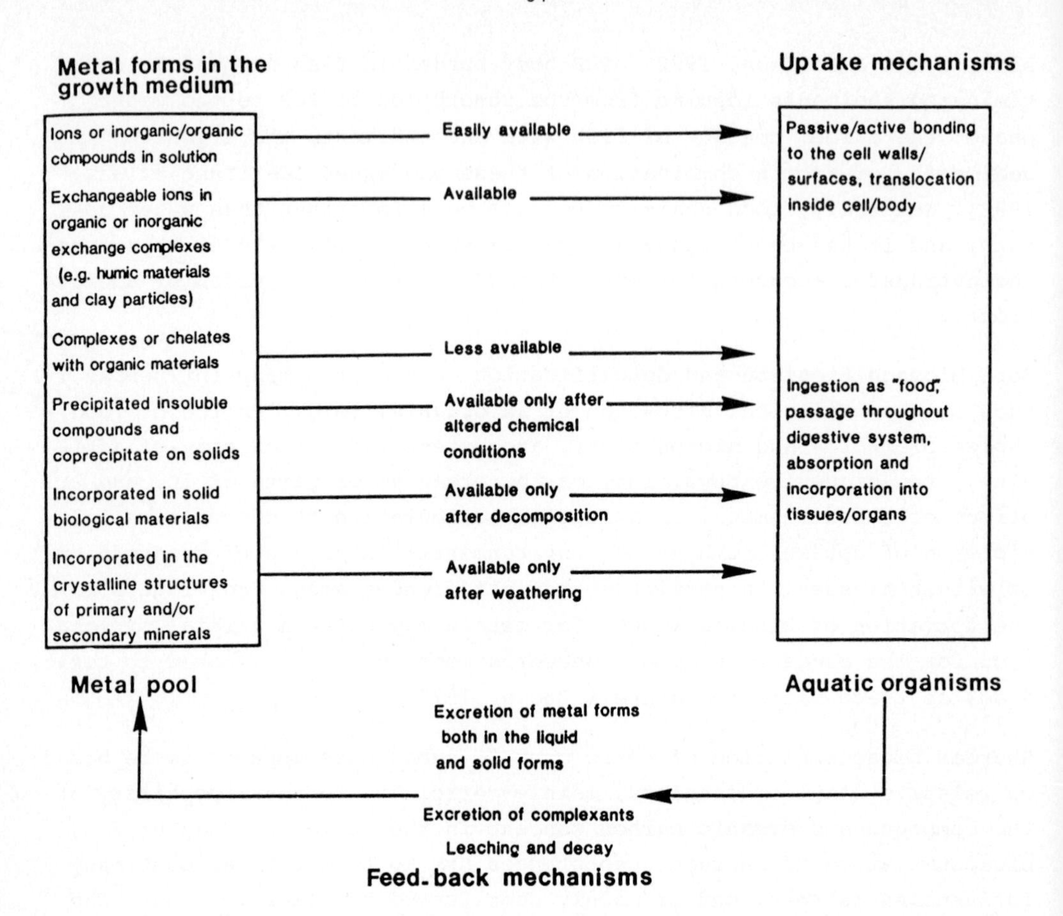

Figure 5-6 Availability of Metal Forms for Biological Uptake (after Baudo, 1982)

leachable fraction does not necessarily correspond to the amount available to biota (Pickering, 1981). This handicap is primarily due to a lack of information about the specific mechanism by which organisms actively translocate trace element species. In the case of **plant root** activities interactions with soil and sediment components include redox changes, pH alterations and organic complexing processes.

Studies on the **prediction** of the trace metal levels in benthic organisms have shown, that the prognostic value of **sequential extraction** data is improved, when the trace metal concentrations are **normalized** with respect to the iron (hydrous oxide) and/or organic content of the sediments (Tessier & Campbell, 1987; see Table 5-2). Depending on the main route for accumulation of trace metals in the benthic organism of concern, at least three mechanisms could explain an inverse relationship between metal concentrations in the organism and the concentration of a potential sink in the sediments:

Table 5-2 Some Examples from *In Situ* Studies for Prediction of Trace Metal Availability to Benthic Organisms from Sediment Characteristics (after Tessier & Campbell, 1987)

Organism	Metal	Best predictor in the sediment	Ref.
Scrobicularia plana	Pb	[Pb]/[Fe] extracted with 1 N HCl	(1)
Scrobicularia plana	As	[As]/[Fe] extracted with 1 N HCl	(2)
Scrobicularia plana and *Macoma balthica*	Hg	[Hg] extracted with HNO_3/organic content (%)	(3)
Anadonta grandis *Elliptio complanata*	Cu	[Cu]/[Fe] extracted with $NH_2OH{\cdot}HCl$	(4) (5)

References: (1) Luoma & Bryan (1978); (2) Langston (1980); (3) Langston (1982); (4) Tessier *et al.* (1983) (5) Tessier *et al.* (1984)

Mechanism 1. If it is assumed that the major route for accumulation of trace metals in the organism involves the **digestive system**, and that **acidic** and/or reducing conditions prevail in the intestinal tract, one can envisage the simultaneous solubilization of both the cations (Fe, Mn, Ca) and the trace metals associated with some sinks (e.g., Fe/Mn oxides). The cations thus released could then compete with the trace metals fro uptake sites in the digestive system and reduce their uptake.

Mechanism 2. If is it again assumed that accumulation of trace metals occurs predominantly via the **digestive system**, but that conditions prevailing within the gut (pH, pE, residence time) are such that the sink remains **unchanged** during the digestion, then the unreacted sink could compete with the uptake sites in the intestinal tract for the solubilized trace metals.

Mechanism 3. Alternatively, the main route for accumulation of the trace metals may involve the uptake of **dissolved** trace metals (e.g., via the gills and mantle). In this case, the protective role of the sink could be explained by invoking adsorption in the external medium as the principal factor controlling the dissolved metal concentrations to which the organisms are exposed. Clearly, as the concentration of adsorbing substrate increases, the concentration of dissolved metals will decrease.

It has been inferred by Tessier & Campbell (1987) that a strong dependence of trace metal accumulation upon sediment characteristics does not imply that the main route of entry of trace metals is necessarily via **ingestion** of particulate metals; it can be explained by a control through adsorption reactions of the dissolved trace metal concentrations in the solution to which the organisms are exposed, such as in the case of **filter-feeders** where high levels of trace metals were found associated with the gills and mantle (Tessier et al., 1984). For the latter pathway the study of the **intermediate water phase** - e.g., pore water - and the different forms and availabilities of metals in this medium seems to be particularly promising (Kersten & Förstner, 1988).

5.5. Case Study: Mobilization of Cadmium from Tidal River Sediments

Within estuaries, intertidal mudflats primarily provide a sink for pollutants imported from upstreams. However, at these productive sites, seasonal effects and even diurnal water level fluctuations induce drastic environmental influences, particularly by redox changes, as has been stressed by Gambrell *et al.* (1977). A case of **"oxidative remobilization"** of trace metals has been described by Kersten & Kerner (1985) from the "Heukenlock", a tidal freshwater flat - forming a 100 ha natural reservate area - in the upper Elbe estuary near Hamburg. Strong enrichment of cadmium in the rhizomes of monodominant reed stands (*Phragmites communis*) colonizing the high flat has indicated high proportions of bioavailable cadmium in the rooting zone of this site. In order to study the chemical forms of cadmium and their potential transfer into water and biota, short (30 cm) sediment cores have been taken from this site and subsamples were analyzed with sequential extraction according to the methods described in the Chapters 3 and 5.

The top 4 cm of the silty core sediments are densily packed with macrophyte litter and reveal a slightly reducing **microhabitat** (Figure 5-7). Below 4 cm the E_h level increases to a plateau at about 300 mV prevailing down to 8-12 cm depth. A second, less well developed redox plateau at about -50 mV occurs from 16 to 20 ± 2 cm depth, followed by a strong anoxic sediment zone. This zonation is obviously controlled by the more or less water-logged conditions in the sediment during the studied period, since the groundwater table was found to alternate between 8 and 22 ± 2 cm below the sediment surface immediately after and before inundation. The decline of NH_4-N in the upper 10 cm reflects the aerobic nitrification during the relative long aeration periods of the sediment surface, whereas rapid transition to lower E_h values in the frequently water-logged horizon indicates the occurrence of anaerobic respiration processes, beginning with denitrification (Figure 5-7). The nitrate level decreases near to zero, whereas the dissolved ammonium concentration is again raised below 12 cm. In the anoxic zone below 18 cm, only a slight depletion in the pore water sulfate concentration was noticed, although a hydrogen sulfide odor occurred. This is probably due to a replenishment of $SO_4{}^{2-}$-supply from the tidal water.

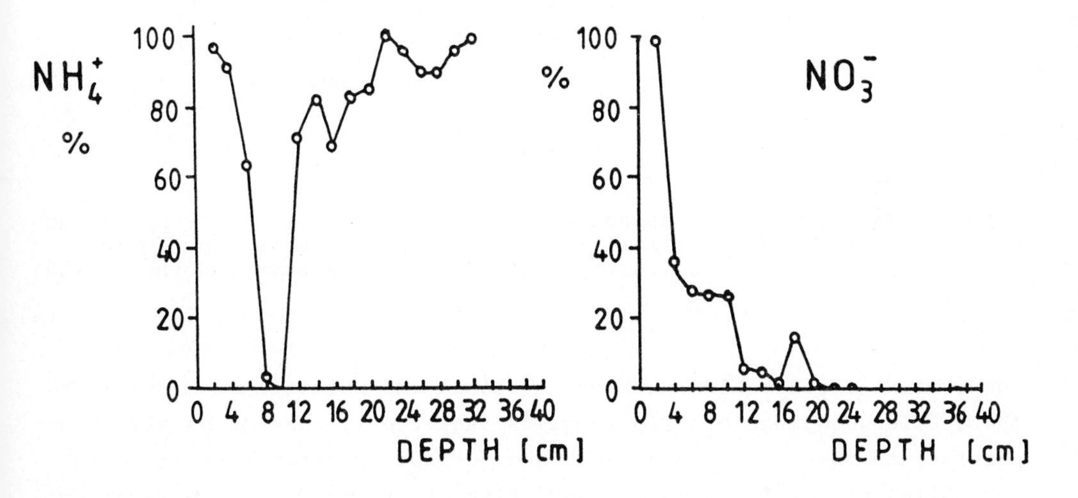

Figure 5-7 Depth Profiles of Nitrogen Compounds in the Heukenlock Area (Kersten & Kerner, 1985)

In the upper part of the sediment column, **total particulate cadmium content** is approximately 10 mg/kg, whereas in the deeper anoxic zone 20 mg Cd/kg have been measured. The results of the sequential extractions of the core sediment samples separated at 2-cm levels (Figure 5-8) indicate, that in the anoxic zone cadmium is associated by 60-80% to the sulfidic/organic fraction. In the oxic and transition zone, sulfidic and organic fractions decrease to approximately 30-40%, whereas carbonatic and exchangeable fractions simultaneously increase up to 40% of total cadmium concentrations. Thus, it is notable that high proportions of mobile cadmium forms correlate with the marked reduction in total cadmium contents.

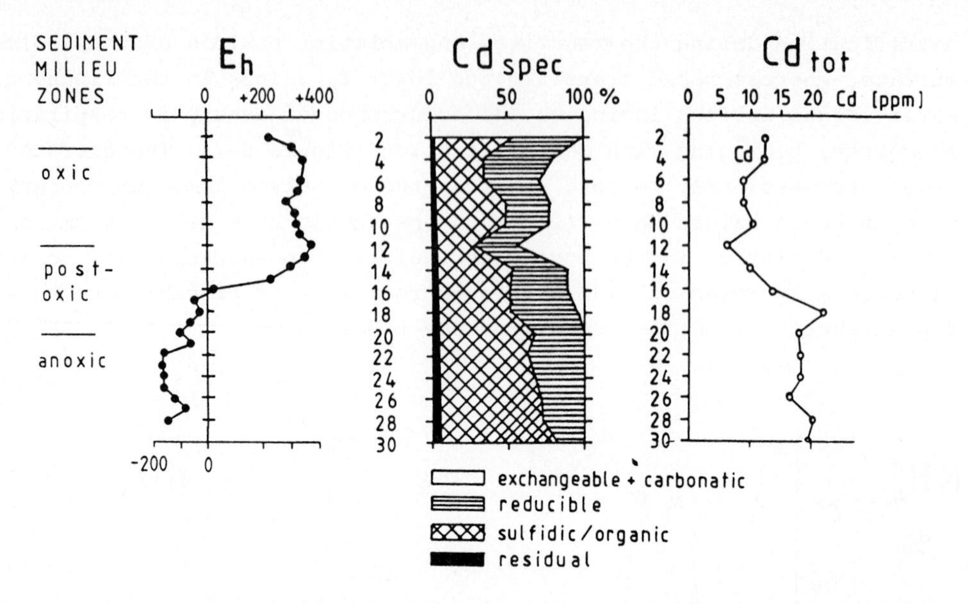

Figure 5-8 Geochemical Characteristics of Core Sediments from the Heukenlock Intertidal Flat, and Total Contents and Chemical Forms of Particulate Cadmium (after Kersten & Kerner, 1985)

The present distribution patterns of total and partitioned cadmium in the studied sediment profile suggests that the release of metals from particulate phases into the water and further transfer into biota is controlled by the frequent **downward flux** of oxidized surface water by tidal action. In the oxic zone, cadmium is leached from the labile particulate-binding sites, where the predominant mechanism controlling the availability of cadmium is adsorption/desorption. With the downward pore water flux, the mobilized metal gets into the anoxic environment, where cadmium is reprecipitated forming sulfidic/organic associations. From the recorded concentrations, it is expected that long-term removal of up to 50% of cadmium from the sediment subsurface will take place at the anoxic sedimentary sink located a few centimeters below the sediment-water interface, which gives a flux maximum of 0.4 g/m^2 per year in the Heukenlock area. The effect of the process of "**oxidative pumping**" (Kersten & Förstner, 1987) on the release of cadmium and other toxic metals into the overlying water and further biogeochemical cycling within the Elbe River estuary will need further investigations.

References to 5.1 "In-Situ Processes"

Allan, R.J. (1986) *The Role of Particulate Matter in the Fate of Contaminants in Aquatic Ecosystems.* National Water Research Institute, Scientific Series No. 142. 128 p. Burlington/Ontario: Canada Centre for Inland Waters.

Baccini, P. (1985) Phosphate interactions at the sediment-water interface. In *Chemical Processes in Lakes*, ed. W. Stumm. New York: Wiley.

Baes, C.F. & Sharp, R.D. (1983) A proposal for estimation of soil leaching and leaching constants for use in assessment models. *J. Environ. Qual.* 12, 17-28.

Batley, G.E. & Giles, M.S. (1980) A solvent displacement technique for the separation of sediment interstitial waters. In *Contaminants and Sediments*, ed. R.A. Baker, Vol. 2, pp. 101-117. Ann Arbor: Ann Arbor Sci. Publ.

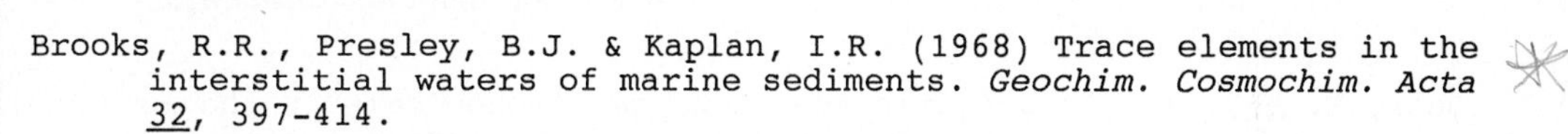
Brooks, R.R., Presley, B.J. & Kaplan, I.R. (1968) Trace elements in the interstitial waters of marine sediments. *Geochim. Cosmochim. Acta* 32, 397-414.

Davison, W. (1985) Conceptual models for transport at a redox boundary, implications for lacustrine geochemistry. In *Chemical Processes in Lakes*, ed. W. Stumm. New York: Wiley.

Duchart, P., Calvert, S.E. & Price, N.B. (1973) Distribution of trace metals in the pore waters of shallow water marine sediments. *Limnol. Oceanogr.* 18, 605-610.

Eadie, B.J. *et al.* (1983) *The Cycling of Toxic Organics in the Great Lakes: A Three-Year Status Report.* NOAA Techn. Mem. ERL GLERL-45. Springfield/VA: NTIS

Elderfield, H. (1981) Metal-organic associations in interstitial waters of Narragansett Bay sediments. *Amer. J. Sci.* 281, 1184-1196.

Förstner, U. & Salomons, W. (1983) Trace element speciation in surface waters: Interactions with particulate matter. In *Trace Element Speciation in Surface Waters and its Ecological Implications*, ed. G.G. Leppard, NATO Conf. Ser. I:6, pp. 245-273. New York: Plenum Press.

Hart, B.T. & Davies, S.H.R. (1977) A new dialysis-ion exchange technique for determining the forms of trace metals in water. *Aust. J. Mar. Freshwater Res.* 28, 105-112.

Honjo, S. (1980) Material fluxes anHonjo, S. (1980) Material fluxes and modes of sedimentation in the mesopelagic and bathypelagic zones. *J. Mar. Res.* 36, 45-57.

Jonasson, I.R. (1977) Geochemistry of sediment/water interactions of metals, including observations on availability. In *The Fluvial Transport of Sediment-Associated Nutrients and Contaminants*, ed. H. Shear & A.E.P. Watson, pp. 255-271. Windsor/Ont.: IJC/PLUARG.

Krom, M.D. & Sholkovitz, E.R. (1978) On the association of iron and manganese with organic matter in anoxic marine pore water. *Geochim. Cosmochim. Acta* 42, 607-611.

Lee, F.Y. & Kittrick, J.A. (1984) Elements associated with the cadmium phase in a harbor sediment as determined with the electron beam microprobe. *J. Environ. Qual.* 13, 337-340.

Lu, C.S.J. & Chen, K.Y. (1977) Migration of trace metals in interfaces of seawater and polluted surficial sediments. *Environ. Sci. Technol.* 11, 174-182.

Luther, G.W. *et al.* (1980) Metal sulfides in estuarine sediments. *J. Sediment. Petrol.* 50, 1117-1120.

Maes, A. & Cremers, A. (1986) Radionuclide sorption in soils and sediments: Oxide-organic matter competition. In *Speciation of Fission and Activation Products in the Environment*, eds. R.A. Bulman & J.R. Cooper, pp. 93-100. London: Elsevier Applied Science Publ.

Nissenbaum, A. & Swaine, D.J. (1976) Organic matter-metal interactions in recent sediments. The role of humic substances. *Geochim. Cosmochim Acta* 40, 809-816.

Petr, T. (1977) Bioturbation and exchange of chemicals in the mud-water interface. In *Interactions between Sediments and Fresh Water*, ed. H.L. Golterman, pp. 216-226. The Hague and Wageningen/The Netherlands: Junk & PUDOC.

Presley, B.J., Kolodny, Y. & Nissenbaum, I.R. (1972) Early diagenesis in a reducing fjord, Saanich Inlet, British Columbia. II. Trace element distribution in interstitial water and sediment. *Geochim. Cosmochim. Acta* 36, 1073-1090.

Salomons, W. (1985) Sediments and water quality. *Environ. Technol. Lett.* 6, 315-368.

Salomons, W. & Baccini, P. (1986) Chemical species and metal transport in lakes. In *The Importance of Chemical "Speciation" in Environmental Processes*, eds. M. Bernhard *et al.*, Dahlem-Konferenzen, pp. 193-216. Berlin: Springer-Verlag.

Salomons, W. *et al.* (1987) Sediments as a source for contaminants? In *Ecological Effects of In-Situ Sediment Contaminants*, eds. R.L. Thomas *et al.* Dordrecht: Dr. W. Junk Publ. (*Hydrobiologia* 149, 13-30.

Schmitt, H.W. & Sticher, H. (1986) Prediction of heavy metal contents and displacement in soils. *Z. Pflanzenernähr. Bodenk.* 149, 157-171.

Van Den Berg, C.M.G. & Dharmvanij, S. (1984) Organic complexation of zinc in estuarine interstitial and surface water samples. *Limnol. Oceanogr.* 29, 1025-1036.

References to 5.2 "Mobilization of Pollutants"

Compeau, G. & Bartha, R. (1983) Effects of sea salt anions on the formation and stability of methylmercury. *Bull. Environ. Contam. Toxicol.* 31, 486-493.

Craig, P.J. (1986) *Organometallic Compounds in the Environment.* Harlow: Longman.

Dickson, W. (1980) Properties of acidified waters. In *Ecological Impact of Acid Precipitation,* eds. D. Drablos & A. Tollan, pp. 75-83. Oslo-Aas/Norway: SNSF Project.

Emerson, S., Jahnke, R. & Heggie, D. (1984) Sediment-water exchange in shallow water estuarine sediments. *J. Mar. Res.* 42, 709-730.

Fagerström, T. & Jernelöv, A. (1972) Aspects of the quantitative ecology of mercury. *Water Res.* 6, 1193-1202.

Förstner, U. (1981) Trace metals in fresh waters (with particular reference to mine effluents). In *Handbook of Strata-Bound and Stratiform Ore Deposits,* ed. K.H. Wolf, Vol. 9, pp. 271-303. Amsterdam: Elsevier Publ. Co.

Gendron, A. *et al.* (1986) Early diagenesis of cadmium and cobalt in sediments of the Laurentian Trough. *Geochim. Cosmochim. Acta* 50, 741-747.

Hines, M.E. *et al.* (1984) Seasonal metal remobilization in the sediments of Great Bay, New Hampshire. *Mar. Chem.* 15, 173-187.

Holmes, C.W., Slade, E.A. & McLerran, C.J. (1974) Migration and redistribution of zinc and cadmium in marine estuarine systems. *Environ. Sci. Technol.* 8, 255-259.

Hunt, C.D. & Smith, D.L. (1983) Remobilization of metals from polluted marine sediments. *Can. J. Fish. Aquat. Sci.* 40, 132-142.

Kersten, M. *et al.* (1985) Freisetzung von Metallen bei der Oxidation von Schlämmen. *Vom Wasser* 65, 21-35.

Lu, C.S.J. & Chen, K.Y. (1977) Migration of trace metals in interfaces of seawater and polluted surficial sediments. *Environ. Sci. Technol.* 11, 174-182.

Morris, A.W., Bale, A.J. & Howland, R.M.J. (1982) The dynamics of estuarine manganese cycling. *Estuar. Coastal Shelf Sci.* 13, 175-192.

Salomons, W. (1983) Trace metals in the Rhine, their past and present (1920-1983) influence on aquatic and terrestrial ecosystems. Proc. Intern. Conf. *Heavy Metals in the Environment, Heidelberg.,* pp. 764-771. Edinburgh: CEP Consultants.

Salomons, W. & Förstner, U. (1984) *Metals in the Hydrocycle.* 349 p. Berlin: Springer-Verlag

Salomons, W. & Mook, W.G. (1980) Biogeochemical processes affecting metal concentrations in lake sediments (IJsselmeer, The Netherlands). *Sci. Total Environ.* 16, 217-229.

Singer, P.C. & Stumm, W. (1970) Acidic mine drainage: The rate-determining step. *Science* 167, 1121-1123.

Summers, A.O. & Silver, S. (1978) Microbial transformations of metals. *Ann. Rev. Microbiol.* 32, 637-672.

Westall, J., Leuenberger, C. & Schwarzenbach, R.P. (1985) Influence of pH and ionic strength on the aqueous-nonaqueous distribution of chlorinated phenols. *Environ. Sci. Technol.* 19, 193-198.

Windom, H.L. (1976) Environmental aspects of dredging in the coastal zone. CRC Critical Reviews in Environmental Control, Vol. 5, pp. 91-109. Boca Raton/FL: CRC Press.

References to 5.3 "Metal Transfer between Inorganic and Organic Substrates"

Ahlf, W., Calmano, W. & Förstner, U. (1986) The effects of sediment-bound heavy metals on algae and importance of salinity. In *Sediments and Water Interactions*, ed. P.G. Sly, pp. 319-324. New York: Springer-Verlag.

Calmano, W., Ahlf, W. & Förstner, U. (1988) Study of metal sorption/-desorption processes on competing sediment components with a multi-chamber device. *Environ. Geol. Water Sci.* 11, 77-84.

Luoma, S.N. & Davis, J.A., 1983. Requirements for modeling trace metal partioning in oxidized estuarine sediments. *Mar. Chem.* 12: 159-181.

References to 5.4 "Transfer to Biota"

Baudo, R. (1982) The role of the speciation in the transfer of heavy metals along the aquatic food web. In *Ecological Effects on Heavy Metals Speciation in Aquatic Ecosystems*, ed. O. Ravera, Ispra-Courses, Manuscript, 31 p. Ispra/Italy: Euratom Centre.

Califano, R.J., O'Connor, J.M. & Hernandez (1982) Polychlorinated biphenyl dynamics in Hudson River striped bass. *Aquat. Toxicol.* 2, 187-204.

Campbell, P.G.C. *et al.* (1988) *Biologically Available Metals in Sediments.* Associate Committee on Scientific Criteria for Environmental Quality Report NRCC No. 27694, 298 p. Ottawa: National Research Council Canada.

Förstner, U. & Wittmann, G.T.W. (1979) *Metal Pollution in the Aquatic Environment.* 489 p. Berlin: Springer-Verlag

Fowler, S.W. *et al.* (1978) Polychlorinated biphenyls: Accumulation from contaminated sediments and water by the polychaete Nereis diversicolor. *Mar. Biol.* 48, 303-309.

Goerke, H. & Ernst, W. (1977) Fate of ^{14}C labelled di-, tri-, and pentachlorobiphenyl in the marine annelid Nereis virens. *Chemosphere* 9, 551-558.

Jenne, E.A. & Luoma, S.N. (1977) Form of trace elements in soils, sediments, and associated waters: An overview of their determination and biological availability. In *Biological Implications of Metals in the Environment*, eds. R.E. Wildung & H. Drucker, Conf.-750929, pp. 110-143. Springfield/Va: NTIS.

Kersten, M. & Förstner, U. (1988) Assessment of metal mobility in dredged material and mine waste by pore water chemistry and solid speciation. In *Chemistry and Biology of Solid Waste - Dredged Material and Mine Tailings*, eds. W. Salomons & U. Förstner, pp. 214-237. Berlin: Springer-Verlag.

Langston, W.J. (1980) Arsenic in U.K. estuarine sediments and its availability to deposit-feeding bivalves. *J. Mar. Biol. Assoc. U.K.* 60, 869-881.

Langston, W.J. (1982) Distribution of mercury in British estuarine sediments and its availability to deposit-feeding bivalves. *J. Mar. Biol. Assoc. U.K.* 62, 667-684.

Luoma, S.M. (1983) Bioavailability of trace metals to aquatic organisms. A review. *Sci. Total Environ.* 28, 1-22.

Luoma, S.M. & Bryan, G.W. (1978) Factors controlling the availability of sediment-bound lead to the estuarine bivalve Scrobicularia plana. *J. Mar. Biol. Assoc. U.K.* 58, 793-802.

Lynch, T.R. & Johnson, H.E. (1982) Availability of a hexachlorobiphenyl isomer to benthic amphipods from experimentally contaminated natural sediments. In *Aquatic Toxicology and Hazard Assessment*, ed. J.G. Person *et al.*, ASTM ST766, pp. 273-287. Washington, D.C.: American Society of Testing and Materials.

Munawar, M. *et al.* (1984) *An Overview of Sediment-Associated Contaminants and their Bioassessment.* Techn. Rept. Can. Fisheries Aquatic Sciences, No. 1253, 136 p. Ottawa: Environment Canada

Nriagu, J.O. (Ed.)(1984) *Changing Metal Cycles and Human Health.* Dahlem-Konferenzen, Life Sciences Research Report 28, 367 p. Berlin: Springer-Verlag.

Pickering, W.F. (1981) Selective chemical extraction of soil components and bound metal species. *CRC Crit. Rev. Anal. Chem.* Nov., 233-266

Reynoldson, T.R. (1987) Interactions between sediment contaminants and benthic organisms. In *Ecological Effects of* In-Situ *Sediment Contaminants*, eds. R.L. Thomas et al., *Hydrobiologia* 149: 53-66.

Salomons, W. & Förstner, U. (1984) *Metals in the Hydrocycle.* 349 p. Berlin: Springer-Verlag.

Smies, M. (1983) Biological aspects of trace element speciation in the aquatic environment. In *Trace Element Speciation in Surface Waters and its Ecological Implications*, ed. G.G. Leppard, pp. 177-193. New York: Plenum Press.

Tessier,A. & Campbell, P.G.C. (1987) Partitioning of trace metals in sediments: Relationships with bioavailability. In *Ecological Effects of* In-Situ *Sediment Contaminants*, eds. R.L. Thomas *et al.*, *Hydrobiologia* 149: 43-52.

Tessier, A., Campbell, P.G.C. & Auclair, J.C. (1983) Relationships between trace metal partitioning in sediments and their bioaccumulation in freshwater pelecypods. In Proc. Intern. Conf. *Heavy Metals in the Environment, Heidelberg,* pp. 1086-1089. Edinburgh: CEP Consultants.

Tessier, A. *et al.* (1984) Relationships between the partitioning of trace metals in sediments and their accumulation in the tissues of the freshwater mollusc Elliptio complanata in a mining area. *Can. J. Fish. Aquat. Sci.* 41, 1463-1472.

Thomann, R.V. & Connolly, J.P. (1984) Model of PCB in the Lake Michigan lake trout food chain. *Environ. Sci. Technol.* 18, 65-71.

References to 5.5 "Case Study: Mobilization of Cadmium from Tidal River Sediments"

Gambrell, R.P. *et al.* (1977) *Trace and Toxic Metal Uptake by Marsh Plants as Affected by Eh, pH, and Salinity.* Techn. Rept. D-77-40, Vicksburg/MS: U.S. Army Engineer Waterways Experiment Station

Kersten, M. & Förstner, U. (1987) Cadmium associations in freshwater and marine sediment. In *Cadmium in the Aquatic Environment*, eds. J.O. Nriagu & J.B. Sprague, pp. 51-88. New York: Wiley

Kersten, M. & Kerner, M. (1985) Transformations of heavy metals and plant nutrients in a tidal freshwater flat sediment of the Elbe estuary as affected by Eh and tidal cycle. In Proc. Int. Conf. *Heavy Metals in the Environment, Athens,* pp. 533-535. Edinburgh: CEP Consultants.

Theme for Further Consideration

"Modeling Contaminant Mobilization and Transport through Sediments"

6. ENVIRONMENTAL IMPACT

6.1. Biological Effects - Bioassays

The most obvious impact of sediment-associated pollutants on aquatic biota is direct **acute toxicity** and there is considerable literature on both laboratory and field effects of toxic substances on marine and freshwater invertebrates (Baker, 1980; Reynoldson, 1987). For example, Warwick (1980) and Wiederholm (1984) observed deformities in chironomid larvae mouthparts at polluted sites of lakes in Canada and Sweden; Milbrink (1983) has shown setal deformities in oligochaetes exposed to high sediment mercury levels. **Indirect effects** resulting from sediment contamination oftenly include changes in benthic invertebrate community structure. For example, Lock *et al.* (1981) evidenced increased growth of bacterial flora and algal cells on oiled substrates and a consequent stimulation of macroinvertebrates. Chapman *et al.* (1982) have shown effects of life history alterations (e.g., impairment of reproduction and age selective toxicity) which have been linked to sediment contaminants.

Generally, it is difficult to establish clear **cause-and-effect relationships** between acute or chronic toxic effects on biota and the occurrence of specific pollutants in sediments. One major limitation is that not all sediment-associated chemicals can presently be identified; thus, unidentified compounds cannot be ruled out as principal etiological factors. Also, those classes of chemicals that are correlated with certain diseases may not be acting individually but through synergistic/antagonistic interactions which may include presently undetected chemicals. Examples have been presented by Malins and colleagues (1984) from polluted sites in Puget Sound, Washington. Positive correlations were found between the frequencies of liver neoplams (e.g., hepatocellular carcinoma) and other liver lesions in English sole (*Parophrys vetulus*) and concentrations of certain aromatic hydrocarbons in sediment; such correlations were not found with chlorinated hydrocarbons. Additional lines of evidence for cause-and-effect relationships include the facts, that (i) the highest prevalences of the lesions were found in the highly contaminated portions of the urban embayments, (ii) stomach contents (consisting mainly of benthic invertebrates) from English sole had concentrations of a number of aromatic hydrocarbons similar to those in the sediment from which the fish were taken, and (iii) similar types of lesions have been reported to occur in laboratory animals exposed to toxic and/or carcinogenic chemicals.

To allow the prediction of potential environmental effects on aquatic organisms different bioassays are required, depending on the specific objective of such assessments, e.g., regarding in-situ sediment contaminants, sediment resuspension, or disposal of dredged materials (Engler, 1980). A bioassay is defined as "any test in which organisms are used to detect or measure the presence or effect of one or more substances or conditions" (Anon., 1973). Since there are different vectors of contaminant uptake by aquatic organisms, predicting the effect of a pollutant in an ecosystem requires a selection of suitable test organisms.

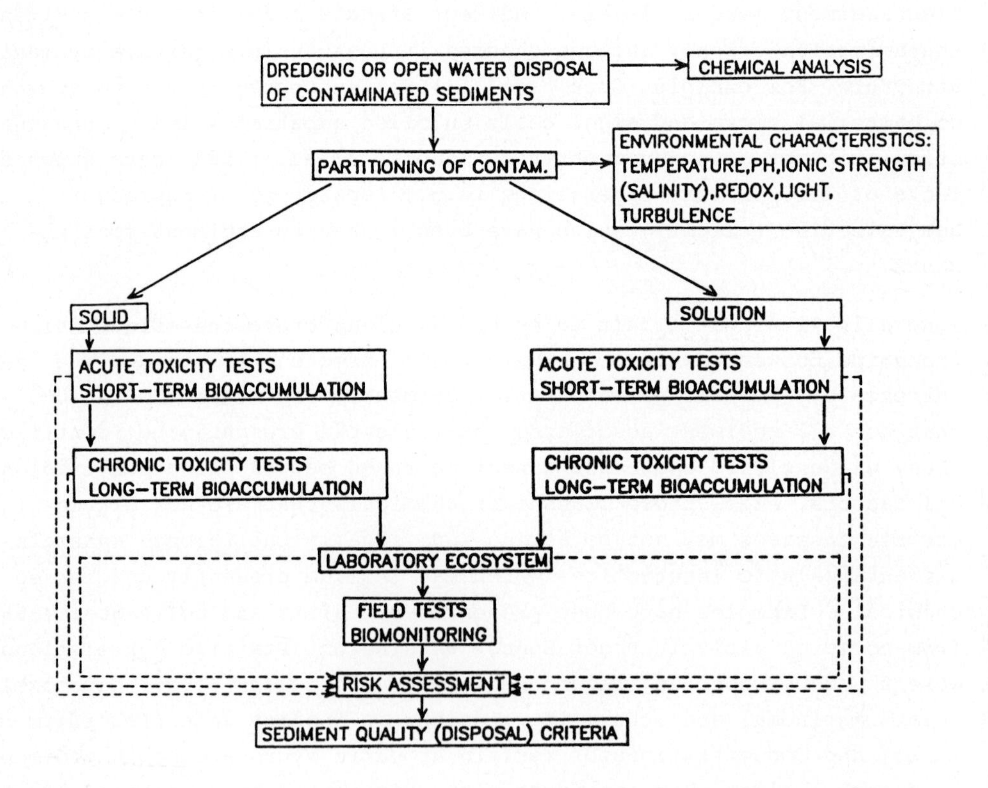

Figure 6-1 Flow Chart for the Assessment of Biological Impact of Sediments and Dredged Material (from Ahlf & Munawar, 1988)

In Figure 6-1 (from Ahlf & Munawar, 1988, after Calamari *et al.*, 1979) the proposed strategy for **risk assessment** of sediment-associated pollutants is shown. The solid lines indicate the direction of increasing difficulty and specificity of each level. The dotted lines show that at each level, a risk assessment is possible when the results of the test represents either a toxic or hazardous bioconcentration of contaminants. Different types of **biological tests** have been applied on polluted sediments, either on liquid-phase or water-column effect, and those concerned with effects of solid-bound contaminants, which are ingested by aquatic organisms:

- **Acute toxicity and short-term bioaccumulation.** Studies of acute toxicity measure the lethal response after 24 or 96 h of exposure in various bodies of water. Test species should be chosen from amongst most commonly used organisms and standardized procedures should be applied. The test should include at least three different trophic levels, namely primary producers, primary consumers and secondary consumers. Normally, organisms such as green algae, daphnids and fish are utilized.

- **Chronic toxicity and long-term bioaccumulation.** The long-term sublethal effects of sediments are measured after at least 96 h of exposure (usually some weeks) using frow-through water. Chronic toxicity tests should be done on the species discussed above. Tests on the reproduction, development and growth of aquatic organisms are commonly used as well as experiments on physiological parameters, taking into consideration levels of organization lower than the whole organism (i.e., organs, enzymes).

- **Laboratory ecosystems.** Multispecies laboratory microcosms are proposed as tools for demonstrating ecological effects both biologically and statistically. Examples for microcosms, the statistical properties of the data and design criteria for microcosms were presented by Crow & Taub (1979).

- **Field tests and biomonitoring.** In-situ test methods verify the findings of laboratory tests in field situations. A major weakness is that this type of evidence has not proven adequate for extrapolation to other ecosystems (Cairns, 1981), but relating field measurements to the effects occuring in laboratory tests may contribute to the interpretation of the bioassay results.

- **Risk assessment and sediment quality criteria.** Risk assessment does not necessarily represent an ordered sequence of elements involving the recognition of hazards, the measurement of impact and the comparison of the measurements. Rather, all possible combinations of elements exist in practice. Sediment quality criteria, among other approaches (see 6.3) use kinetic bioaccumulation models and bioaccumulation tests (Chapman *et al.*, 1987)

In general, the bioassay results must be interpreted in relation to the field situations and site-specific data. In section 6.5, a conceptional model "Sediment Triad" (Chapman, 1986), which combines data from bulk sediment chemistry, bioassay and *in-situ* studies, is presented.

6.2. Pollution Indices

A quantitative measure of metal pollution in aquatic sediments has been introduced by Müller (1979), which is called the "**Index of Geoaccumulation**":

$$I_{geo} = \log_2 C_n / 1.5 \times B_n$$

C_n is the measured concentration of the element "n" in the pelitic sediment fraction (<2 μm) and B_n is the geochemical background value in fossil argillaceous sediment ("average shale"); the factor "1.5" is used because of possible variations of the background data due to lithogenic effects. The Index of Geoaccumulation consists of 7 grades, whereby the highest grade (6) reflects 100-fold enrichment above the background values ($2^6 = 64 \cdot 1,5$). In Table 6-1 an example is given for the River Rhine; and a comparison of these sediment indices with the water quality classification of the International Association of Waterworks in the Rhine Catchment (IAWR) has been made. The index values for the upper Rhine (between Basel and Mannheim, before the waters of the highly polluted Neckar and Main Rivers enter the Rhine) lie at a maximum of 3 for Cd and Pb. Cadmium reaches the highest grade (6) in the middle section of the river and the lower Rhine, which can be trace to influents from the Neckar River and from the Ruhr area. The indices for Pb and Zn attain grade 4, and those for Hg, Cu and Cr each increase by one grade compared to the upper Rhine River section. The detailed situation of the I_{geo} indices in longitudinal profiles of the Rhine River is shown in Figure 6-2 (after Müller, 1979).

Table 6-1 Comparison of IAWR Water Quality Indices (Based on Biochemical Data) and Index of Geoaccumulation (I_{geo}) of Trace Metals in Sediments of the Rhine River (after Müller, 1979)

IAWR Index	IAWR water quality (pollution intensity)	Sediment accumulation (I_{geo})	I_{geo}-class	Metal examples Upper Rhine	Metal examples Lower Rhine
4	very strong pollution	>5	**6**		Cd
3-4	strong to very strong	>4-5	**5**		
3	strongly polluted	>3-4	**4**		Pb, Zn
2-3	moderately to strongly	>2-3	**3**	Cd, Pb	Hg
2	moderately polluted	>1-2	**2**	Zn, Hg	Cu
1-2	unpolluted to mod.poll.	>0-1	**1**	Cu	Cr, Co
1	practically unpolluted	<0	**0**	Cr, Co	

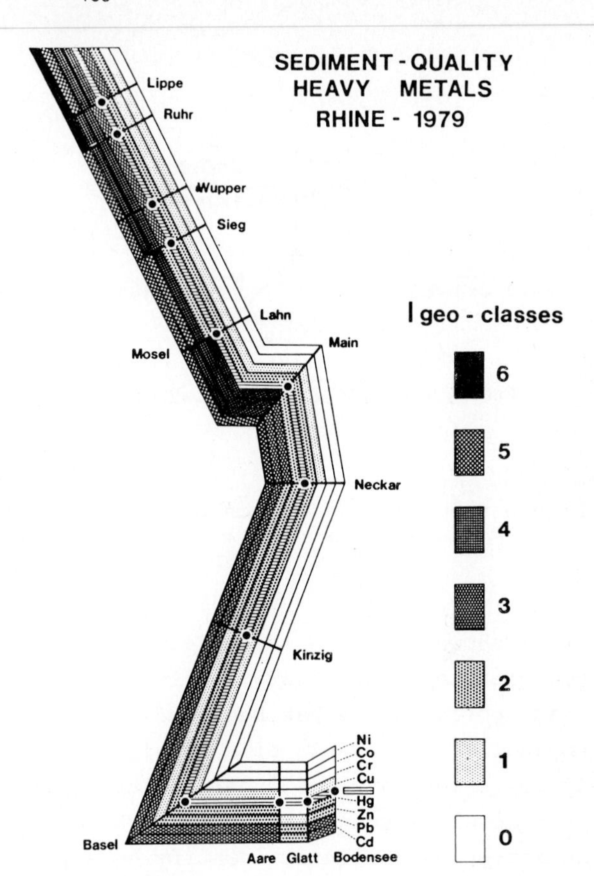

Figure 6-2

Index of Geoaccumulation in Sediments of the Rhine River (Müller, 1979)

In accordance with a classification of metal levels in estuaries based on enrichment factors in intertidal algae or fauna by Tomlinson *et al.* (1980) a pollution index can be developed for sediment-bound contaminants from sites via zones to whole estuaries or other ecosystems. The **Pollution Load Index** (PLI) is at first determined for the different metals at each **site** by calculating highest contamination factors (C.F. = metal concentration in polluted sediment/base value for that metal is calculated, deriving the n-th root of the n factors multiplied together: n Cf1·Cf2... Cfn). Such site indices can be treated in the same manner to give a **zone index**, and an **estuary index** from the zone indices (Figure 6-3a). Figure 6-3b illustrates such indices using *Fucus sp.* calculated for the east coast estuaries of Ireland (Tomlinson *et al.*, 1980).

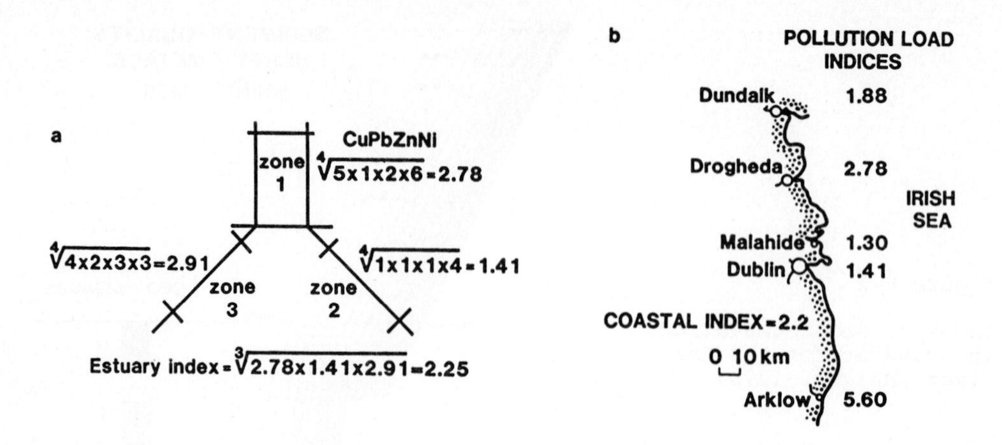

Figure 6-3 Classification of Pollution Load Index (PLI) After Tomlinson *et al.* (1980) **(a)** and Example for the East Coast of Ireland **(b)**. Explanation See Text.

It is shown that the Avoca Estuary at Arklow is by far the most polluted with regard to heavy metals, whereas Malahide and Dublin are the least polluted. It has been inferred by the authors of this study that unusually high levels of a single pollutant might result in an unusually low index level of other pollutants in the system are present at near baseline concentrations. Such faults might be overcome by using a subscript to the index identifying any pollutant present in abnormally high concentrations; thus the index for the Avoca Estuary at Arklow would become 5.6 Cu Zn, indicating that copper and zinc were the major pollutants to give rise to the relatively high Pollution Load Index. For the application of this method on **sediment parameters** it will be necessary to determine an enrichment factor corrected for grain size on the basis of conservative elements (Chapter 3.3).

A sedimentological approach for an **"ecological risk index"** was introduced by Hakanson (1980) and tested on 15 Swedish lakes representing a wide range in term of size, pollution status, trophic status, etc. These estimations are based on four "requirements", which are determined in a relatively rapid, inexpensive and standardized manner from a limited number of sediment samples:

(a) **The concentration requirement** compares data from the uppermost layers with a standardized set of natural background levels of the respective pollutant.

(b) **The number requirement** is built on the idea that for practical purposes only a limited number of substances can be used, which are determined from sediment samples as "total concentrations". In practice, the sum of selected contaminants and their concentration factors (PCB, Hg, Cd, As, Cu, Pb, Cr and Zn) represent the "standard degree of contamination".

(c) **The toxic requirement** differentiates the various contaminants according to an "abundance principle", i.e. saying that there exists a proportionality between toxicity and rarity, and to their "sink-effect", i.e. their affinity to solid substrates. After a normalization process the **"sedimentological toxic factor"** is calculated in the following sequence: Zn = 1 < Cr = 2 < Cu = Pb = 5 < As = 10 < Cd = 30 < Hg = PCB = 40.

(d) **The sensitivity requirement** indicates that different lakes/-basins have different sensitities to the different toxic substances. In this context the **"bioproduction index"** as determined from the data on nitrogen content and loss of ignition is used as the critical parameter for the "toxic-response factor" of a certain contaminant.

To quantitatively express the potential ecological risk of a given contaminant in a given sedimentary environment, the risk factor E_r is defined as $E_r = T_r \times C_r$, where T_r = the toxic response factor for a given substance, and C_r = the contamination factor. Five ranges from low to very high describe the potential **ecological risk factor** (E_r) for a given contaminant; the four-step potential **ecological risk index** (RI) is the sum of the individual risk factors, i.e. of the selected contaminants, for a certain lake/basin. The examples from Swedish lakes in Table 6-2 show that within the different classes of risk indices (RI-values) there may be quite characteristic **"contamination profiles"** with prevalences for certain pollutants. For example, Lake Norra Barken is characterized by high potential ecological risk factors originating from elevated contents of PCB and Cd, while in Lake Varmlandssjon - in the same group of very high risk indices - the concentrations of Hg pose the predominant problems. It has been stressed by Hakanson (1980) that this approach could at first indicate the overall pollution impact in a certain environment, expressed by the RI-status, but then - even more important in the practice of pollution control - the risk factors point to those substances to which the most attention should be paid.

6–2 The potential ecological risk factors (Er′-values) and risk indices (RI-values) of the 15 Swedish lakes. All figures marked * are hypothetical

Lake	RI	Potential ecological risk factor (E_r')				
		$E_r' > 320$	$320 > E_r' > 160$	$160 > E_r' > 80$	Moderate $80 > E_r' > 40$	Low $E_r' < 40$
Risk index (RI) very high RI > 600						
Vasman	1,201	Hg	Cd	PCB* > Pb	As*	Cu > Zn > Cr
Norra Barken	813		PCB* > Cd	Hg	Pb > As*	Zn > Cu > Cr
Ovre Hillen	741	Hg	PCB*	Cd	As*	Pb > Cu > Zn > Cr
Varmlandssjon	652	Hg			Cd > PCB*	As* > Pb > Cu > Zn > Cr
Stora Aspen	650		PCB*	Cd > As*	Hg > Pb > Cr	Zn > Cu
Risk index (RI) considerable 300 < RI < 600						
Vanern	511	Hg			Cd > PCB*	As* > Pb > Cu > Zn > Cr
Osterjon	415			PCB* > Hg	Cd > As*	Cr > Cu > Pb > Zn
Hjalmaren	380		PCB*		Cd*	Hg* > As* > Pb > Cu > Cr > Zn
Amanningen	370		PCB*		Cd > Hg > As*	Pb > Cr > Zn > Cu
Freden	365		PCB*	Cd	Hg	As* > Cr > Pb = Cu = Zn
Malaren	326			PCB* > Hg	Cd	As* > Cu > Pb > Cr > Zn
Blacken	305		PCB*		Cd > Hg	As* > Cu > Cr > Pb > Zn
Risk index (RI) moderate 150 < RI < 300						
Haggen	275			Cd	PCB* > Hg	As* > Pb > Cu > Zn > Cr
Vattern	248			PCB*	Cd* > Hg	As* > Pb > Cu > Zn > Cr
Bysjon	231			PCB*	Hg > Cd	As* > Pb > Cu = Zn > Cr
Risk index (RI) low RI < 150						

6.3. Long-Term Prognosis on Sediment-Induced Groundwater Quality

Predictive models are limited due to the difficulties to describe sorption kinetics between (persistent) chemicals and solid phases. In such cases it could be advantageous for directly measuring changes of chemical forms of typical soil and waste constituents at in-situ conditions with respect to interstitial water composition, e.g. by **inserting dialysis bags** containing typical substrates with varying metal dosages. Figure 6-4 shows initial results on the behaviour of low metal-dosage to iron oxyhydrates and solid organic substrates under anaerobic conditions. In both examples lead is released during the 4-weeks experiment particularly from easily reducible phases. For copper slight increase is observed during the experimental period, again in the easily reducible fraction. With the **direct method** relationships between anthropogenic substances and waste/natural soil substrates can be evaluated for:

(i) **site-specific** properties of different disposal systems, which can be compared with respect to their compatibility with the waste material, consisting of a wide range of pollutants;

(ii) influences of a **stabilized system** of hydrological and chemical factors, mechanisms and processes - including biochemical interactions - on the behaviour of individual critical pollutants.

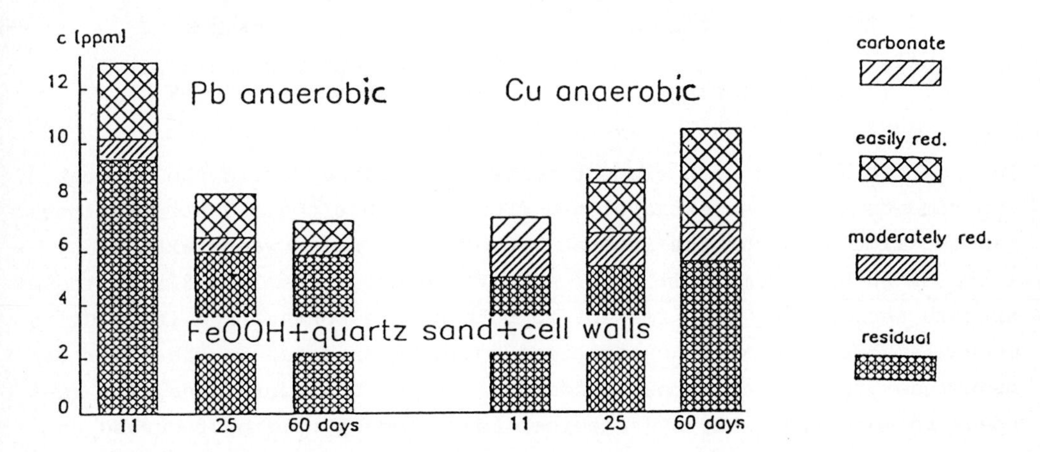

Figure 6-4 Changes of Total Concentration and Chemical Forms of Pb and Cu on Inorganic and Organic Substrates Inserted into Anaerobic Groundwater (Förstner & Carstens, 1988)

In Chapter 3.4 a method has been presented for long-term prognosis of metal pollutant mobility, which combines **column circulation** leaching experiments at variable pH/E_h-conditions with **sequential extraction** procedures on the solid waste material *before* and *after* these experiments (Schoer & Förstner, 1987). Temporal release patterns are different for the individual elements (Figure 6-5): While at pH 5/400 mV release of cadmium seems to be completed within the experimental period mobilization of copper is still going on and the end point cannot be estimated from the data of the "kinetic" experiments. The same effect has been found for the examples of thallium and vanadium. For the other elements, the endpoint of release can be determined as approximately 10 mg cobalt, 0.6 mg cadmium, 600 mg zinc and 0.3 mg chromium, 2 mg barium and 20 mg lead (per 100 g of solid substrate treated with 140 L solution.

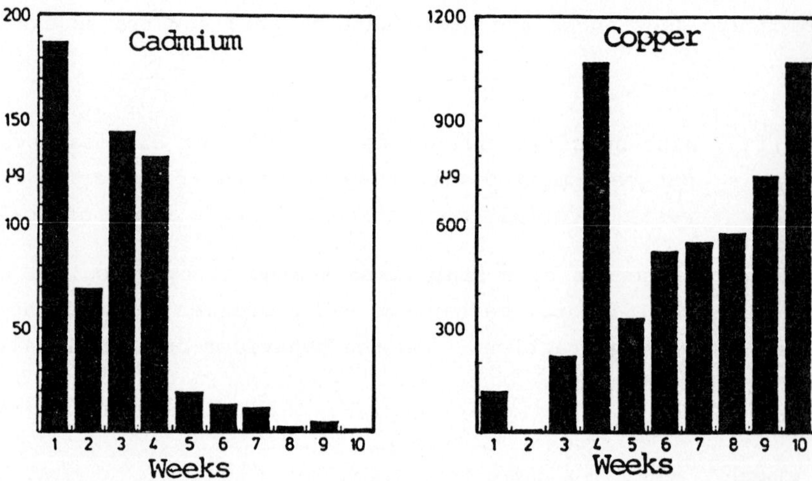

Figure 6-5 Temporal Evolution of Cadmium and Copper Release (in µg). 100 g Sample with 140 L Solution at pH 5/+ 400 mV

Taking into account both element contents released during the 10 week experiments and those extrapolated from reducible pools concentrations can be calculated for the interaction of 100 g solid waste with 140 litre of pH 5/ 400 mV (Tl: pH 8/ 400 mV) solution (Table 6-3). At these extreme assumptions with respect to both solute contact and interactive mechanisms most metal concentrations would be expected in the order of magnitude of drinking water standards. Further considerations with respect to more realistic hydrological conditions should be focussed on the examples of zinc and lead, where excessive concentrations could be derived from the present model (Table 6-3).

Table 6-3 Calculation of Dissolved Metal Concentrations in Leachates from Heat-Processing Waste Materials from Direct Determinations (Treatment of 100 g Sample with 140 Litre Extraction Solution) and from Estimations of Mobilizable Metal Pools.

Mobilization at Flow-Through Conditions*			Mobilization-Relevant Metal-"Pool" in mg/kg				Results in Diss. Concentr. mg/L	Water Quality Criteria mg/L
	per kg	pH/Redox	Acetate (1)	Oxalate (2a)	(2b)	H_2O_2 (3)		
Cd	6 mg	pH 5/ 400	[1		2	1]	0.004	0.005
Co	100 mg	pH 5/ 400	[6	90]	40	5	0.07-0.1	0.05
Cr	3 mg	pH 5/ 400	0		250	15	0.002	0.01[a]
Cu	50 mg	pH 5/ 400	[15	35]	185	50	0.03-0.20	0.10
Pb	600 mg	pH 5/ 400	[20		350	200]	0.40	0.05
Tl	20 mg	pH 8/ 400	[10	10]	30	5	0.01-0.03	0.01
Zn	6000 mg	pH 5/ 400	[1000		5000]	300	3.0	1.0

*140 L solution/100g solid waste   [a] value for Cr(VI)

Differences in the temporal development of release rates for the individual elements are connected with their sorption/desorption behaviour, which is primarily due to **pH-effects** but may also be influenced by complexation, e.g. by elevated concentrations of chloride ions. With respect to the pH-effects, however, there are significant differences in the response of the various solid substrates to the addition of H^+-ions, and it may be argued that the pH-values on the solid surfaces - which can be estimated from "pH-titration tests" - are decisive for the behavior of the particular element rather than the pH-values determined in solution.

The combination of **mobilisation kinetics** and the chemical **metal forms** enable to prognosticate the long-term release behaviour of metals from the landfill under estimation. For this purpose the "quick-motion" factor of the experimental design has to be evaluated. For zinc under the flow-through conditions at pH 5 and 400 mV, from the total release (437 mg per kg) and the amount of water passing through the landfill material (140 l), an experimental zinc release of approx. 3 mg/l can be calculated (Table 6-3). The measured concentrations in the pore-water of the deposition site are between 0.04 mg Zn/l and 1.40 mg Zn/l, in the groundwater just beside the landfill approx. 0.02 mg Zn/l. If all this zinc in the nearby groundwater would have been resulted from the landfill (worst case assumption), a "black-box-experimental-factor" of 150 can be calculated from the ratio of the proposed and the real zinc concentrations. If this factor is multiplied with the duration of the experiment (10 weeks), a time period of nearly 30 years is represented by the experiment. From Figure 3-7 it can be concluded, that the release of the rest of "oxalate-bound" zinc would take another 12 experimental weeks; this would be equivalent to nearly 40 years in nature. Altogether, a time period of at least 50 to 100 years would be covered by the experiment, and a prognosis within this time scale can be drawn from the present results.

The "quick-motion-factor" of 150 represents a minimum, because it is based on the assumption that all zinc measured in the groundwater in the immediate neighbourhood of the landfill is release from that site. However, similar concentrations are very common in groundwater aquifers not affected significantly by anthropogenic activities. If significant amounts of zinc are not originating from the deposit, the experimental factor is higher and the time scale of prognosis, therefore, would be even longer. In any case, after this time period the zinc concentration in groundwater will decrease. The upper **limits of the prognosis** can be evaluated by the proton balance because the pH-value is the most important parameter for the metal release in many such examples. The porewater of the waste material exhibits a pH-value of 6, i.e. the proton concentration is 10^{-3} mmol/l. During the experiment, conducted during a time period of 10 weeks, 25 mmol H^{+}/l have been added to maintain pH 5, equivalent to a time factor of 25.000; this results in an extrapolated time period of 5.000 years, represented by the experiments on the basis of the proton balance. For the present example, however, even such mobilisation cannot be expected in reality, since there is no internal process - such as sulfide oxidation or acetate production from organic degradation - which could affect pH conditions in the deposits. Additional calculations on the basis of atmospheric inputs of acidity can be interpreted in a way that with the present form of deposition of the metal oxide residues from heat processing below the groundwater level and excluding degradable organic substances no significant change of metal mobility or release will take place with several thousand years.

Long-term evolution of **reactive materials**, involving decomposition of organic substances, is still an open question. Oxidation of sulfidic minerals by intruding rainwater may mobilize trace metals, and the impact on the underlying groundwater could be even higher if a chromatography-like process of continuous dissolution and reprecipitation would preconcentrate such chemicals prior to final release with the leachate.

6.4. Sediment Quality Criteria

Apart from the assessment and prognosis of effects of sediment-associated pollutants on benthic organisms (6.1.) and potential hazards on groundwater quality (6.3), three major reasons have been given for the establishment of sediment quality criteria:

- In contrast to the strong temporal and spatial variability in the aqueous concentrations of contaminants, sediments integrate contaminant concentrations over time, and can, therefore, reduce the **number of samples** in monitoring, surveillance and survey activities;
- long-term perspectives in water resource management involve **"integrated strategies"**, in which sediment-associated pollutants have to be considered;
- waste water plans will increasingly be based on the **assimilative capacity**[1] of a certain receiving system, which requires knowledge of properties of sedimentary components as the major sink.

Efforts have been undertaken mainly be the United States Environmental Protection Agency to develop standard procedures and criteria for the assessment of environmental impact of sediment-associated pollutants. Initial discussions (Anon., 1984, 1985) suggested **five methodological approaches** which merit closer consideration: (i) "background approach", (ii) "water quality/pore water approach", (iii) sediment/water equilibrium partitioning approach", (iv) "sediment/organism equilibrium" approach, and (v) "bioassay" approach. Of these possibilities, applications of "bioassays" and "background approach" have been outlined in sections 6.1 and 6.2, respectively.

---

[1] The term "assimilative capacity" was introduced by a group of marine scientists and engineers (Goldberg, 1979) to define the amount of material that could be contained within a body of seawater without producing an unacceptable biological impact. This amount, essentially determined by a titration of the pollutant with the water body, becomes evident at an "endpoint". The most extensive set of endpoints has evolved because of artificial radioactive nuclides entering the oceans from nuclear fuel cycles; the impact of DDT and its degradation products upon non-target organisms such as fish-eating birds could have provided endpoints for its presence in the oceans. Pollutant concentrations that are determined *before* the endpoint is reached are "checkpoints". While the role of sediments in estimating the assimilative capacity has not explicitely been mentioned in either one of the site-specific considerations (Deepwater Dumpsite 106; New York Bight; Southern California Bight; Puget Sound), there is an intial attempt for calculating cadmium concentrations in organisms from equilibrium partition coefficients to sediments (Atwood *et al.*, 1979)

A schematic diagram in Figure 6-6 indicates the various procedures used for estimation the hazard of contaminants associated with sediments (Chapman *et al.*, 1987):

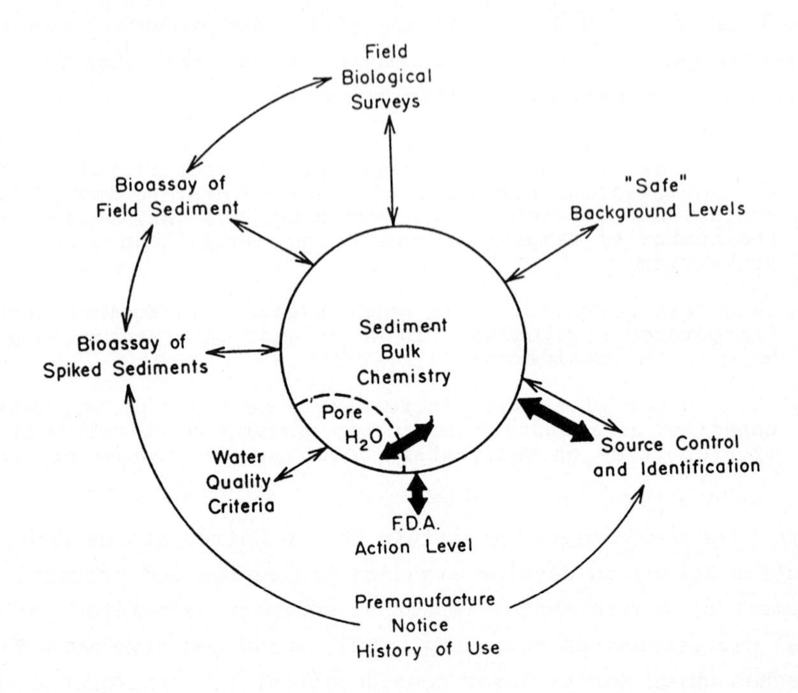

Figure 6-6 Schematic Diagram of the Various Procedures Used for Estimating the Hazard of Contaminants Associated with Sediments (G. Chapman *et al.*, 1987). Dark Arrows Represent Modeling Applications.

- **Background levels** - compare site sediment chemistry with that of other sites determined to be unimpacted (numerical pass/fail)

- **Field biological surveys** - conduct on-site studies of biota to evaluate possible impact at site (biological response pass/fail)

- **Bioassay of spiked sediment** - estimate effect/no effect sediment concentration for a specific chemical (numerical criterion). May be desired in clearance of new chemicals.

- **Tissue action level** - link sediment concentration to safe tissue concentration (e.g., FDA action level or body burden-response data) through application of equilibrium or kinetic models (numerical criterion).

- **Aqueous toxicity data** - apply toxicity data from typical water-column bioassays to sediments through direct measurement of pore water concentration or estimation of pore water from sediment concentrations through application of equilibrium models. May be desired in evaluation of new chemicals.

Biological Criteria

Biological criteria exhibit major advantages in that they integrate effects of multiple factors including sediment characteristics and complex or unknown wastes, and, with respect to field surveys, that they are site-specific, requiring minimum extrapolations. On the other hand, field surveys are costly and bioassay organisms may not represent sensitivity of the natural species assemblage (Anderson *et al.*, 1987).

Relatively simple and implementable liquid, suspended particulate and solid-phase bioassays have been carried out for assessing the short-term impact of dredging and disposal operations on aquatic organisms (sections 6.1, 7.2 and 7.3). Standardized tests are characterized by their lack of variability, but essential information (e.g., lethality, alterations of growth rate) can only be obtained with such single species test. The influence of the main environmental variables on the interaction of suspended particulates or in-situ sediment contaminants and organisms should also be determined under simulated field conditions. Nonetheless, a great deal of research was done to find suitable organisms for bioassays: For example, natural phytoplankton size assemblages have been used successfully in the North American Great Lakes (Munawar & Munawar, 1987), and benthic bioassay procedures, due to recent developments, are important in evaluating the relationship between laboratory and field impacts (Reynoldson, 1987).

Numerical Criteria

Major advantages of numerical criteria lie in their easy application and amendment to **modeling approaches**. However, if criteria do not exist for certain chemicals of concern, a biological test may still be required. In addition some equilibrium modeling approaches will fail if tissue concentration and toxicity is independent (Chapman *et al.*, 1987).

Sediment/water equilibrium partitioning. This approach is related to a relative broad toxicological basis of water quality data. The distribution coefficient K_D, which is determined from laboratory experiments, is defined as the quotient of equilibrium concentration of a certain compound in sediment (C_s^x, e.g. in mg/kg) and in the aqueous phase (C_w^x; e.g. in mg/l). In practice, three categories of compounds can be distinguished:

a) **Nonpolar organic compounds**, which are dominantly correlated to the content of organic carbon in the sediment sample. The partition coefficient K_D can be normalized from this parameter and the octanol/water-coefficient (K_{OW}): $K_D = 0.63\ K_{OW}$/content of organic carbon in total dry sediment (0.63 is an empirical value). For these substances, such as PCB, DDT, and PAH reliable and applicable data can be expected with respect to the development of sediment quality criteria.

b) K_D-values of **metals** are not only correlated to organic substances but also with other sorption-active surfaces. Toxicological effects oftenly are inversely correlated with parameters such as iron oxyhydrate (Chapter 5.3). Quantification of competing effects is difficult, and thus the equilibrium partition approach for sediment quality assessment of metals still exhibits strong limitations.

c) **Polar organic substances** (e.g. phenols, polymers with functional groups, tensids) are widely unexperienced with respect to their specific "sorption" behaviour. Partition coefficients are influenced by anion and cation exchange capacity and surface charge density as a function of pH and other complex properties, so that the K_D-approach at present cannot be taken into consideration.

For a number of nonpolar hydrophobic organic contaminants, in particular for critical PAH compounds (section 6.1), **interim sediment criteria** values have been proposed by U.S. Environmental Protection Agency (Anon., 1988).

Sediment/biota equilibrium partitioning. A very important aspect of the assessment of the environmental fate of chemicals is the prediction of the extent to which these substances will achieve concentrations in biotic phases. For organic chemicals, it has been suggested by Mackay (1982) that the **bioconcentration factor K_B** can be regarded simply as a partition coefficient between an organism consisting of a multiphase system and water; if the dominant concentrating phase is a lipid that has similar solute interaction characteristic to octanol, a proportional relationship between bioconcentration factor K_B and $K_{octanol/water}$ is expected ($K_B = 0.048\ K_{OW}$). This correlation must be used with discretion, particularly for very low K_B, where the amount of solute in non-lipid phases may be appreciable, and for high-K_{OW} compounds (e.g.

mirex, octachlorosterene, and higher chlorinated biphenyls). In fact, Oliver (1984) found characteristic dependencies of bioconcentration of oligochaete worms in sediments from K_{OW}-values of the chemicals, where a slow linear increase in bioconcentration factors with K_{OW} is observed for chemicals with K_{OW}'s less than 10^5, and a rapid decrease in bioconcentration occurs for chemicals with very high partition coefficients ($>10^6$). The decrease may be caused by difficulties in chemical transport across worm membranes due to large molecular size or may be affected by strong binding of these chemicals to the sediments making them less bioavailable (see Chapter 5).

Similar to the aqeuous equilibria this approach is based on a relative broad experience with food quality data. Here too, the three categories of compounds - non-polar organic substances, metals, and polar organic substances - can be distinguished; in practice, again, only for the non-polar organic compounds sufficient experience is available for developing quality criteria from equilibrium data. As outlined, there is a strong correlation for hydrophobic organic substances between the bioconcentration factor K_B and the octanol/water coefficient, indicating that the lipid content of the organism constitutes the major concentrating phase in the system water/organism. Simple **steady-state correlative models** have been used in laboratory studies to predict the bioconcentration from water of organic compounds by fishes, mussels and other aquatic organisms. One major difficulty is that these models were developed from steady-state concentrations and laboratory systems lacking suspended particulate material, and thus cannot deal with the varying concentrations and forms of pollutants found in many environments (Lake *et al.*, 1987). Inclusion of **solid particulate matter** considers bioaccumulation as a redistribution of contaminants between sources (organic carbon of waste materials) and sinks (dissolved phase and lipids of organisms). For conditions in which the aqueous phase is not important as a sink - for high solid/aqueous partition coefficients - the bioaccumulation factor will depend on the concentration in the source (organic carbon) and sink (lipids of organisms). Partition factors (PFs) between sediments and organisms have been defined as (Lake *et al.*, 1987):

$$PF = \frac{C^i/\text{g sediment (dry wt)/g oc/g sediment (dry wt)}}{C^i/\text{g organism (dry wt)/g lipid/g organism (dry wt)}}$$

PF-values for chlorinated compounds from experiments where exposure concentrations were constant and could be established (i.e., concentrations in sediment were used for infauna; concentrations in suspended particulate matter in dosing systems were used for mussels) were similar to the partition factor of approximately 0.5, which has been calculated by McFarland (1984) from $K_{oc}/BCF_{(lipid)}$ under various assumptions. These findings indicate that modeling bioaccumulation as a redistribution of contaminants between organic carbon of sediments and lipids of organisms is justified for at least come nonpolar chlorinated organics, organisms, and exposures.

Prediction of **bioavailable heavy metal** concentration appears to be more complex, and appropriate normalizing factors still have to be evaluated. Until predictive methods for determining bioavailability of contaminants in sediments can be validated, empirical measurements of body burden and effects as determined by the toxicity test and field monitoring provide the most direct approach for evaluating the impact of contaminated sediment in the aquatic environment (Fava *et al.*, 1987).

6.5. Case Study: Sediment Quality Triad on Puget Sound, Washington

The **sediment quality triad** (Long & Chapman, 1985) combines chemistry and sediment bioassay measures with *in-situ* studies (Figure 6-7): Chemistry and bioassay estimates are based on laboratory measurements with field collected sediments. *In-situ* studies may include, but are not limited to, measures of resident organism histopathology, benthic community structure and bioaccumulation/metabolism. Areas where the three facets of the trad show the greatest overlap (in terms of positive or negative results) provide the strongest data for determing **numerical sediment criteria.**

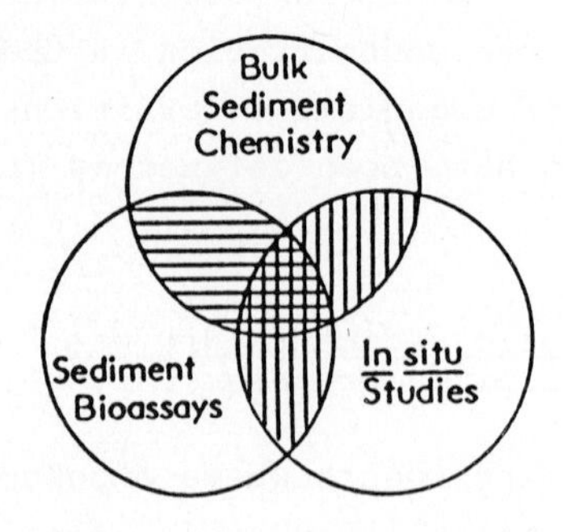

Figure 6-7

Conceptual Model of the Sediment Quality Triad, Which Combines Data from Chemistry, Bioassay and *In-Situ* Studies (Chapman, 1986)

The Puget Sound region receives some of the wastes from a population of about 2 million. Although many areas of Puget Sound are still considered to be relatively pristine, the estern shore is heavily industrialized (Figure 6-8). Tacoma has a larger **copper smelter** and several chemical plants, Seattle has a number of heavy industries and manufacturing plants, and Everett and Tacoma have **wood products industries**. The pulp mill activities have provided an evident source of pollutants. Commencement Bay (Tacoma) has been designated by the U.S. Environmental Protection Agency as on of the top 10 worst **hazardous waste sites** in the U.S. (Malins *et al.*, 1984).

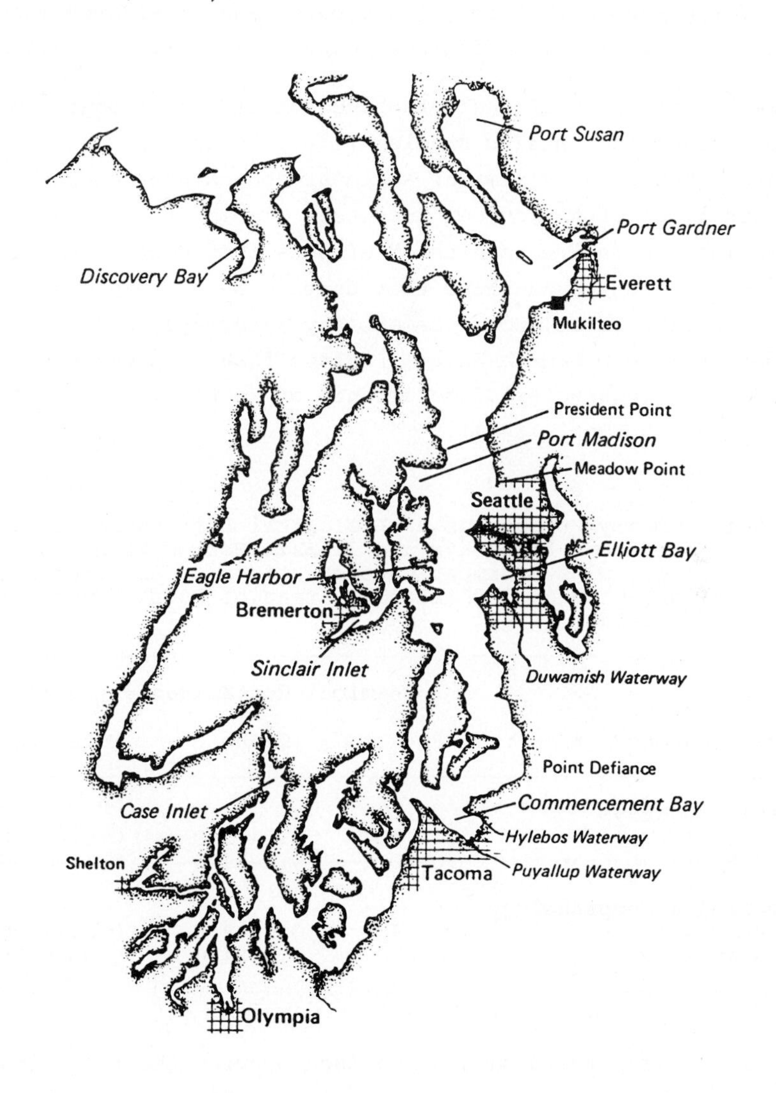

Figure 6-8 Map of Puget Sound, Washington (after Malins *et al.*, 1987)

In the latter area and in Waterways sediment three dominant and representative **chemical groups** were distinguished in the analyses and were selected for further study: high molecular weight combustion polyaromatic hydrocarbons (CPAHs), polychlorinated biphenyls (PCBs) and lead.

Three types of sediment **bioassay** were considered: the amphipod *Rhepoxynius abronius* acute lethality test, the oligochaete *Monopylephorus cuticulatus* respiration effects test, and the fish cell anaphase aberration test. Bottom **fish histopathology** was based on the frequency of selected liver lesions in English sole (*Parophrys vetulus*) from different areas of Puget sound; liver lesions have been considered most likely to be related to chemical contaminant exposures (Malins *et al.*, 1984).

A summary comparison of the data based on **effects frequencies** of sediment bioassays and *in-situ* studies (i.e., bottom fish histopathology) indicates that roughly similar sediment contaminant concentrations produce both types of biological responses (Table 6-4). **Bioassay data** are divided into those frequencies of effects that contain all rural areas (e.g., Case Inlet), and those that contain only urban, industrialized areas (Sinclair Inlet, Commencement Bay Waterways, and Elliott Bay Waterfront and the Duwamish River). **Bottom fish histopathology data** are divided into frequencies of occurrence of up to 5% and those that are greater than 5% (Table 6-4).

Table 6-4 Summary Comparison of Biological Effects Frequencies (Sediment Bioassays and *In-Situ* Bottom Fish Histopathology) with Sediment Concentrations of Selected Chemical Contaminants in Different Areas of Puget Sound, Wash. (Chapman, 1986).

	Chemical Contaminants (μg/g)		
Effects frequency (%)	Pb	CPAHs	Total PCBs
Sediment bioassays			
15 - 50 %	20- 50	0.2- 5.0	0.01-0.10
55 - 80 %	90-800	3.8-24.0	0.10-0.90
Bottom fish histopathology			
0 - 5 %	20- 90	0.2- 3.8	0.01-0.20
6 - 40 %	130-800	6.8-24.0	0.80-0.90

The present **macro-scale data comparison** provide the following useful information about the study area (Chapman, 1986):

(a) Chemical **concentrations** below which biological effects levels are **low** or minimal, (b) chemical concentrations above which biological effects levels are always **high**, and (c) intermediate chemical concentrations representing an area of uncertainty or a **"break-point"** between the high and low concentrations (Table 6-5).

Table 6-5 Sediment Quality Criteria for Lead, Combustion Polyaromatic Hydrocarbons and Total Polychlorinated Biphenyls, as Derived from Data in Table 6-4 (after Chapman, 1986)

Criteria description	Pb	CPAHs	Total PCBs
No or minimal biological effect	≤50	≤3.8	≤0.1
Major biological effects	≥130	≥6.8	≥0.8
Area of uncertainty	>50<130	>3.8<6.8	>0.1<0.8

Derivation of these concentrations ignores which particular contaminants may be causing the observed biological effects. On the other hand, this treatment may provide more realistic **site-specific criteria** than single-chemical characterizations such as are undertaken in other approaches, e.g., determination of partition coefficients (sect. 6.4).

References to 6.1 "Biological Effects - Bioassays"

Ahlf, W. & Munawar, M. (1988) Biological assessment of environmental impact of dredged material. In *Chemistry and Biology of Solid Waste - Dredged Material and Mine Tailings*, eds. W. Salomons & U. Förstner, pp. 127-142. Berlin: Springer-Verlag.

Anonymus (1973) *Biological Field and Laboratory Methods for Measuring the Quality of Surface Waters and Effluents*. U.S. EPA Report No. 670/4-73-001. Cincinnati: United States Environmental Protection Agency.

Baker, R.A. (Ed.)(1980) *Contaminants and Sediments*. 2 Vols., Ann Arbor: Ann Arbor Sci. Publ.

Calamari, D., Galassi, S. & Da Gasso, R. (1979) A system of tests for the assessment of toxic effect on aquatic life: An experimental preliminary approach. *Ecotoxicol. Environ. Saf.* 3, 75-89.

Cairns, J. Jr. (1981) Biological monitoring. Part VI - Future needs. *Water Res.* 15, 941-952.

Chapman, G. *et al.* (1987) Regulatory implications of contaminants associated with sediments. In *Fate and Effects of Sediment-Bound Chemicals in Aquatic Systems*, eds. K.L. Dickson *et al.*, pp. 413-425. New York: Pergamon Press.

Chapman, P.M. (1986) Sediment quality criteria from the sediment quality triad: An example. *Environ. Toxicol. Chem.* 5, 957-964.

Chapman, P.M. *et al.* (1982) *Survey of Biological Effects of Toxicants upon Puget Sound Biota. I. Broad Scale Toxicity Survey.* NOAA Techn. Mem. OMPA-25, 98 pp. Boulder/CO: United States Natl. Oceanic and Athmospheric Administration.

Crow, M.E. & Taub, F.B. (1979) Designing a microcosm bioassay to detect ecosystem level effects. *Int. J. Environ. Studies* 13, 141-147.

Engler, R.M. (1980) Prediction of pollution potential through geochemical and biological procedures: Development of regulatory guidelines and criteria for the discharge of dredged fill material. In *Contaminants and Sediments*, ed. R.A. Baker, Vol. 1, pp. 143-169. Ann Arbor: Ann Arbor Sci. Publ.

Lock, M.A. *et al.* (1982) The effects of synthetic crude oil on microbial and macroinvertebrate benthic river communities. *Environ. Pollut.* 24, 207-217.

Malins, D.C. *et al.* (1984) Chemical pollutants in sediments and diseases of bottom-dwelling fish in Puget Sound, Washington. *Environ. Sci. Technol.* 18, 705-713.

Mibrink, G. (1983) Characteristic deformities in tubificid oligachaetes inhabiting polluted bays of Lake Vänern, Southern Sweden. *Hydrobiologia* 106, 169-184.

Reynoldson, T.B. (1987) Interactions between sediment contaminants and benthic organisms. In *Ecological Effects of* In Situ *Sediment Contaminants*, eds. R.L. Thomas *et al.* *Hydrobiologia* 149, 53-66.

Warwick, W.F. (1980) Palaeolimnology of the Bay of Quinte, Lake Ontario, 2800 years of cultural influence. *Can. Bull. Fish. Aquat. Sci.* 206, 1-117.

Wiederholm, T. (1984) Incidence of deformed chironomid larvae (Diptera: Chironomidae) in Swedish lakes. *Hydrobiologia* 109, 243-249.

References to 6.2 "Pollution Indices"

Hakanson, L. (1980) An ecological risk index for aquatic pollution control. A sedimentological approach. *Water Res.* 14, 975-1001.

Müller, G. (1979) Schwermetalle in den Sedimenten des Rheins - Veränderungen seit 1971. *Umschau* 79, 778-783.

Tomlinson, D.L. *et al.* (1980) Problems in the assessment of heavy-metal levels in estuaries and the formation of a pollution index. *Helgol. Meeresunters.* 33, 566-575.

References to 6.3 "Prognosis on Sediment-Induced Groundwater Quality"

Förstner, U. & Carstens, A. (1988) In-Situ-Versuche zur Veränderung von festen Schwermetallphasen in aeroben und anaeroben Grundwasserleitern. *Vom Wasser* 71, 113-123.

Schoer, J. & Förstner, U. (1987) Abschätzung der Langzeitbelastung von Grundwasser durch die Ablagerung metallhaltiger Feststoffe. *Vom Wasser* 69, 23-32.

References to 6.4 "Sediment Quality Criteria"

Anderson, J. *et al.* (1987) Biological effects, bioaccumulation, and ecotoxicolgy of sediment-associated chemicals. In *Fate and Effects of Sediment-Bound Chemicals in Aquatic Systems*, eds. K.L. Dickson, A.W. Maki & W.A. Brungs, pp. 267-295. New York: Pergamon Press.

Anonymus (1984) *Background and Review Document on the Development of Sediment Criteria.* EPA Contract No. 68-01-6388. McLean/VA: JRB Associates.

Anonymus (1985) *Sediment Quality Criteria Development Workshop*, Nov. 28-30, 25 p. Richland: Battelle Washington Operations.

Anonymus (1988) *Interim Sediment Criteria Values for Nonpolar Hydrophobic Organic Contaminants.* U.S. EPA Office of Water Regulations and Standards, Criteria and Standards Division. SCD#17, May 1988, 34 p. Washington D.C.: U.S. Environmental Protection Agency.

Atwood, D. *et al.* (1979). The New York Bight. In *Assimilative Capacity of U.S. Coastal Waters for Pollutants*, ed. E.D. Goldberg. Proc. Workshop held at Crystal Mountain, Washington, July 29 - August 4, 1979, pp. 148-178. Boulder/CO: U.S. Department of Commerce.

Chapman, G. *et al.* (1987) Regulatory implications of contaminants associated with sediments. In *Fate and Effects of Sediment-Bound Chemicals in Aquatic Systems*, eds. K.L. Dickson, A.W. Maki & W.A. Brungs, pp. 413-425. New York: Pergamon Press.

Farrington, J.W. & Westall, J. (1986) Organic chemical pollutants in the oceans and groundwater: a review of fundamental chemical properties and biogeochemistry. In *Role of the Ocean as a Waste Disposal Option*, ed. G. Kullenberg, pp. 361-425. Dordrecht: D. Reidel Publ.

Fava, J. *et al.* (1987) Workshop summary, conclusions and recommendations. In *Fate and Effects of Sediment-Bound Chemicals in Aquatic Systems*, eds. K.L. Dickson, A.W. Maki & W.A. Brungs, pp. 429-445. New York: Pergamon Press.

Goldberg, E.D. (Ed.) (1979) *Assimilative Capacity of U.S. Coastal Waters for Pollutants*. Proc. Workshop held at Crystal Mountain, Washington, July 29 - August 4, 1979, 284 p. Boulder/CO: NOAA/-U.S. Department of Commerce.

Lake, J.L., Rubinstein, N. & Pavignano, S. (1987) Predicting bioaccumulation: Development of a simple partitioning model for use as a screening tool for regulating ocean disposal of wastes. *Fate and Effects of Sediment-Bound Chemicals in Aquatic Systems*, eds. K.L. Dickson, A.W. Maki & W.A. Brungs, pp. 151-166. New York: Pergamon Press.

Mackay, D. (1982) Correlation of bioconcentration factors. *Environ. Sci. Technol.* 16, 274-278.

McFarland, V.A. (1984) Activity-based evaluation of potential bioaccumulation for sediments. In *Dredging and Dredged Material Disposal*, eds. R.L. Montgomery & J.W. Leach, Vol. 1, pp. 461-467. New York: American Soc. of Civil Engineers.

Munawar, M. & Munawar, I.F. (1987) Phytoplankton bioasseays for evaluating toxicity of in-situ sediment contaminants. In *Ecological Effects of* In Situ *Sediment Contaminants*, eds. R.L. Thomas *et al. Hydrobiologia* 149, 87-105.

Oliver, B.G. (1984) Uptake of chlorinated contaminants from anthropogenically contaminated sediments by oligochaete worms. *Can J. Fish. Aquat. Sci.* 41, 878-883.

Reynoldson, T.B. (1987) Interactions between sediment contaminants and benthic organisms. In *Ecological Effects of* In Situ *Sediment Contaminants*, eds. R.L. Thomas *et al. Hydrobiologia* 149, 53-66.

References to 6.5 "Case Study: Sediment Quality Triad on Puget Sound"

Chapman, P.M. (1986) Sediment quality criteria from the sediment quality triad: An example. *Environ. Tox. Chem.* 5, 957-964.

Long, E.R. & Chapman, P.M. (1985) A sediment quality triad: Measures of sediment contamination, toxicity and infaunal community composition in Puget Sound. *Mar. Pollut. Bull.* 16, 405-415.

Malins, D.C. *et al.* (1984) Chemical pollutants in sediments and diseases of bottom-dwelling fish in Puget Sound, Washington. *Environ. Sci. Technol.* 18, 705-713.

Malins, D.C. *et al.* (1987) Sediment-associated contaminants and liver diseases in bottom-dwelling fish. In *Ecological Effects of* In-Situ *Sediment Contaminants*, eds. R.L. Thomas *et al. Hydrobiologia* 149: 67-74

Theme for Further Consideration

"Evaluation of Baseline Data and Implications Using the 'Background Approach'"

7. DREDGED MATERIALS

7.1. Introduction

As shipping demands minimal water-action or current within the harbour basins, this means that optimal conditions have been created for the sedimentation of river or sea-borne material. In order to keep these ports and channels accessible to (marine) shipping, this material has to be removed regularly by dredging. The International Association of Ports and Harbors (IAPH 1981) received 108 responses from 37 countries on a questionnaire of the year 1979: There were 350 million tonnes of **maintenance dredging** and 230 mill. tonnes average annual **new construction dredging**; this survey found that about one-fourth of all dredged material is ocean-dumped and another two-thirds is deposited in wetlands and nearshore. In the river mouths to the southern coast of the North Sea, for example, approx. 20 Million m^3 have to be dredged from Rhine/Meuse (Rotterdam Hr.) and approx. 10 Million m^3 from the rivers Scheldt (Antwerp), Weser (Bremerhaven) and Elbe (Hamburg) (d'Angremond *et al.*, 1978). The total quantity of sediment which is dredged in The Netherlands, Belgium and West Germany amounts to about twelve times the total suspended matter supply from the Rhine (Van Driel *et al.*, 1984).

Table 7-1 Mean Annual Maintenance Dredging of Harbours Along the Coast of the Southern North Sea (Salomons & Eysink, 1981)

Harbour	Mean annual maintencance dredging in million m^3 per year
Dutch Harbours at the Western Scheldt	2.5
Antwerp	10
Rotterdam	21
Scheveningen	0.2
IJmuiden	2.5
Emden/Delfzijl	15
Bremen/Bremerhaven	10
Hamburg/Cuxhaven	11
London	0.7
Hull	4.8

The possibilities of disposal of these enormous quantities of material are severely limited because of the pollutants present in the dredged material. An example is given in Figure 7-1: In the **Rotterdam** harbor - which will be presented as a case study in section 7.4 - the amount which has to be dredged annually increased from less than 1 Million m^3 in 1920 to approximately 23 Million m^3 in the 1980's. In the last 80 years the cadmium concentrations increased by a factor of approx. 100 (note logarithmic scale in Figure 7-1!).

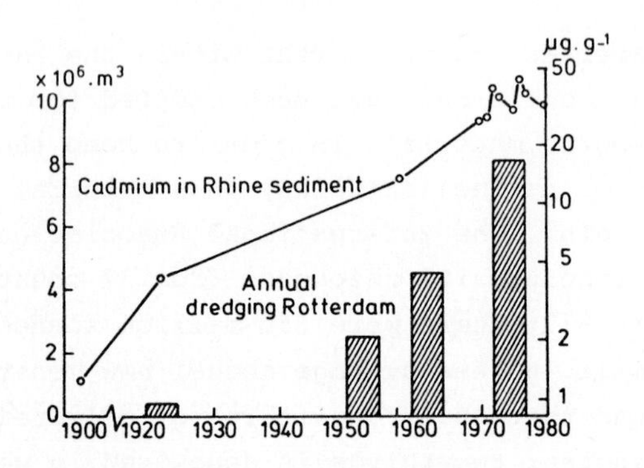

Figure 7-1

Increase in Cadmium Concentrations in Rhine Sediments due to Increased Industrial Use and the Increase in Annual Dredging in the Rotterdam Harbour Due to Harbour Extension (Salomons & Förstner, 1980)

Among the authorities particularly dealing with the subject of contaminants in dredged materials the U.S. Army Corps of Engineer Waterways Experiment Station at Vicksburg, Mississippi has played a leading role (Brannon *et al.*, 1976; Lee *et al.*, 1976)[1]. The Environmental Laboratory of this institution initiated the "Decision-Making Framework for Management of Dredged Material Disposal" (Lee & Peddicord, 1988), which includes **test procedures** on physicochemical conditions, aquatic bioaccumulation, and water column effects at the site of dredging operations as well as mobility determinations (effluent, surface runoff, and leachate quality; plant and animal uptake) at the disposal site.

[1] The latter organisation together with the U.S. Environmental Protection Agency has developed a "Standard Elutriate Test" for detecting the release of chemical contaminants from dredged material (Lee & Plumb (1974). The test involves the mixing of one volume of the sediment to be dredged with four volumes of the disposal site water for a 30-minute shaking period. If the soluble chemical constituent in the elutriate exceeds 1.5 times the ambient concentration in the disposal site water, special conditions will govern the disposal of this dredged material.

7.2. Environmental Impact of Dredging Operations

The physical, chemical and biological effects of dredging operations have been reviewed by Sweeney *et al.* (1975), Sly (1977), Gambrell *et al.* (1977) and Hebert & Schwartz (1983). **Physical effects** are undoubtedly significant at a local level; however, the scale of events, e.g. in Great Lakes, remains small in comparison to sediment resuspension resulting from wind-wave action. **Chemical changes** in dredged material pH and redox potential may occur in interstitial water components and solids when mixed with oxygenated surface water for extended periods. The oxidation of reduced material was found to cause about 12% of oxygen depletion in the central basin of Lake Erie (Burns & Ross, 1972). Investigations by Windom (1975) on the exchange of metals between sediment and water in dredge spoil slurries from estuaries and marsh areas of the southeastern United States did not show significant increases of heavy metals such as lead, copper, iron, and mercury above ambient levels. Generally, metal levels have been found to be fairly constant after an initial re-equilibrium with the sediments. It was suggested by Windom (1976) that even if some metals become more soluble under aerobic conditions, as soon as the sediments are dispersed, reoxidiation would lead to their reprecipitation. Similar indications have been reported from lake sediments in Japan (Otsuka *et al.*, 1976). From studies on Lake Erie sediments, Sly (1977) observed that there is a general increase for manganese in the pore waters after dredging, whereas the highest Fe concentrations appear during dredging. His conclusion were that the early release of heavy metals in soluble form from pore waters is of little concern since in most cases concentrations are within the limits set for potable water and, in any case, the scavenging effect of oxides and hydroxides may rapidly coprecipitate phosphorus and other elements in the water mass soon after disposal. On the other hand, biochemical conversions of certain heavy metals provide a mechanism for the long-term release of toxic species which may be incorporated into the biota by both respiration and body metabolism (see below). These processes, therefore, would merit further consideration.

It has been suggested by Weber *et al.* (1982) that changes in redox potential have a greater effect on the mobility of toxic metals than changes of pH during dredging, while the opposite is true for land disposal (see section 7.3). **Early diagenetic sediment changes** and element mobilization from porewater in a man-made estuarine marsh has been investigated by Darby *et al.* (1986). This study exemplifies both the mechanisms of release of metals via porewater extraction and subsequent changes by the effect of oxidation (Table 7-2): Compared to the river water concentration, the channel sediment porewater is enriched by a factor of 200 for Fe and Mn, 30-50 for Ni and Pb, approx. 10 for Cd and Hg, and 2-3 for Cu and Zn. When the expected concentration of metals following hydraulic dredging, which were calculated from a rate of porewater to river water of about 1:4, were compared with the actual measurements at the pipe exiting the dredging device, charactistic differences were observed (Table 7-2).

Table 7-2 Mobilization of Metals and Nutrients During Dredging (after Darby *et al.*, 1986). All Concentrations in mg/l Except Hg.

Metal	Channel sediment porewater (a)	River water concn. (b)	*a/b*	Effluent at man-made marsh Expected concn.	Measured concn.	**% Change**
Mn	6.94	0.03	*230*	1.34	1.19	**- 11**
Fe	57.3	0.26	*220*	11.12	6.01	**- 46**
Ni	0.054	0.001	*54*	0.011	0.035	**+ 218**
Pb	0.077	0.002	*38*	0.016	0.142	**+ 788**
Hg(µg/l)	3.2	0.26	*12*	0.82	2.0	**+ 144**
Cd	0.009	0.001	*9*	0.0025	0.019	**+ 660**
Cu	0.012	0.004	*3*	0.0055	0.051	**+ 827**
Zn	0.12	0.052	*2*	0.065	5.30	**+ 8069**

When the actual concentration at the pipe's exit were less than expected, that element was removed from solution presumably by scavenging or precipitation; this was valid for iron (approx. half of the expected concentration) and - to a lesser extent - for manganese. When the exiting solution was greater than the expected concentration in Table 7-3, mobilization from sediments was assumed; in this respect, highest rates of release were found for zinc, followed by copper, lead and cadmium.

These data demonstrate the problematic effect of dispersing **anoxic waste materials** in ecologically productive, high-energy nearshore, estuarine, and inlet zones (Khalid, 1980). This may also pertain to procedures such as "sludge-harrowing" as it is occassionally performed in the cold season in some sections of Hamburg harbor. By application of these techniques, highly contaminated sediments are transferred into zone of lower pollution; oxygen-consuming substances, such as ammonia, are released from the pore water; increased turbidity affects "light climate" and thus the ecosystem in the lower reaches of the estuary.

Biological factors in respect to dredging operations include species transformations of toxic metals by methylation (Saxena & Howard, 1977), biodegradation of organic chemicals (Clark *et al.*, 1979), and a wide range of processes affecting bioconcentration of contaminants (Allan, 1986). There are many indications that dredging has immediate, but no long-term effect on benthic organisms (Sweeney *et al.*, 1975). Seeyle *et al.* (1982) found effects of dredging on the accumulation of PCBs, zinc, and mercury in fish. Consequently, it is stressed, that the safety of the ecosystem must be ensured by carefully designed and monitored pilot studies before dredging operations (Hebert & Schwartz, 1983).

7.3. Disposal of Dredged Materials

The major disposal alternatives for dredged sediments and their modifications are listed in Table 7-3 (Gambrell *et al.*, 1978): These **categories** differ primarily in the biological population exposed to the contaminated sediments, oxidation-reduction conditions, and transport pro-

Table 7-2 Disposal Alternatives for Dredged Materials (After Gambrell *et al.*, 1978)

Upland disposal

- Long-term confinement (Ponded/no drainage; non-ponded/dewatering)
- Interim confinement (dewatering and consolidation prior to transport)
- Unconfined upland*
- Habitat development*
- Agricultural soil amendment*

* for less contaminated materials

Intertidal Sites

- Unconfined (mudflats, marsh)
- Confined by boundary structures, e.g. for habitats

Subaqueous Disposal

- Unconfined disposal
- Confined deposition (mounded deposits or capped borrow pits)

cesses potentially capable of removing contaminants from dredged materials at the disposal site. **Grain size**, in particular, of sediment is important because it determines the conditions under which sediment will be resuspended or deposited, it determines the basic habitat available for benthic organisms, and it also determines the surface area to volume ratio of the solid phase which is important in chemical exchange processes with the aqueous phase. A review on the decisionmaking framework for the various disposal options of dredged material has been given by Lee & Peddicord (1988); implications for marine disposal have been discussed by Kester *et al.* (1983).

Disposal on Land and in Shallow Water Areas

One major adverse effect of dredged materials after land disposal relates to **contamination of crops**. Factors controlling plant availability are the nature and properties of the contaminant, of the soil/dredged material, and of the plant species. Predominant substrate factors are texture, concentration and nature of organic matter and pH. High pH, high clay and organic matter contents reduce the plant availability of most metals (Van Driel & Nijssen, 1988). Dredged materials often have high silt and organic matter contents than the corresponding soils, and this may result in lower initial bioavailabilities of the contaminants. In the long run, however, the organic matter will be partly decomposed with a corresponding increase in bioavailability. Another major difference between dredged materials and soils usually is the higher content of **oxidizable sulfide compounds** such as iron sulfide in the former substrate. This results in problems with less buffered dredged sediment.

For example, in the **Hamburg harbour area** where contaminated mud is still pumped into large polders for sedimentation, agriculture is no more permitted at these sites. Due to the low carbonate content, metals are easily transferred to crops during lowering of pH (see below) and permissible limits of cadmium have been surpassed in as much as 50% of wheat crops grown on these materials (Herms & Tent, 1982). This effect can be explained by the ability of certain bacteria (*Thiobacillus thiooxidans* and *T. ferrooxidans*) to oxidize sulfur and ferrous iron; while decreasing the pH from 4-5 to about 2.0, the process of metal dissolution from dredged sludge is enhanced. High concentrations of metals have been measured in oxidized pore waters from sedimentation polders in the Hamburg harbour area (Maaß *et al.*, 1985).

Plants differ in their **ability to accumulate** heavy metals from dredged material (see review Van Driel & Nijssen, 1988). Monocotyledonous plant species accumulate less metals than dicotyledonous species (Cottenie, 1981), but difference between species, varieties and subspecies may mask these differences. Moreover, growth conditions, nutrient status, temperature and transpiration rate determine growth rate and element uptake. The organs of the plant also show different abilities to accumulate metals: for most plant species seeds and fruits accumulate less metal than leaves and roots. **Plant bioassays** have been developed at the U.S. Army Waterways Experiment Station, Vicksburg/Miss. (Folsom *et al.*, 1981); for fresh water conditions mostly the marsh plant *Cyperus esculentus* is used, which may be cultivated in an aerobic or anaerobic rooting environment under upland and flooded growing conditions, respectively; for salt water marsh conditions *Spartina alterniflora* and some other salt water marsh plant species are applied.

Particularly for less buffered soil/dredged material systems, **liming** may be effective in the depression of metal transfer into plants. In calcareous river sediments, e.g. dredged material from the Rotterdam harbor basins (see section 7.5), liming apparently has no effect (Van Driel & Nijssen, 1988). The use of a **soil cover layer** system generally seems to be more effective. Model calculations show that in the case of a soil cover of a least 1.0 m there is no net transfer of dissolved contaminants from the contaminated soil to the clean soil cover. A prerequisite is that the groundwater level is fixed at 1 m below the surface. Experiments with plants show a decrease in the accumulation of metals by the plants, but roots of deep-rooting plants can still reach the contaminated soil layer and contribute to the metal status of the above-ground plant organs (Van Driel, 1985).

With respect to the transfer of pollutants from shallow water sediments into the ecosphere it has been shown that one of the major mechanism of **pollutant release** is oxidation of organic and sulfidic particulate matter, subsequent to mechanical or biological turbation of bottom sediments (Chapter 5.2). On the other hand, it has been suggested that following oxidation of the surface sediment the ecosystem is rapidly recovering and that an oxidized, bioturbated surface layer constitutes an effective barrier against the transfer of most trace metals from below into the overlying water (Salomons *et al.*, 1987). Nonetheless it is indicated that dispersing contaminated sediments generally is much more problematic than "containing" them (Förstner et al., 1986).

Incorporation in naturally formed minerals, which remain stable over geological times, constitutes favourable conditions for the **immobilization** of potentially toxic metals in large-volume waste materials both under environmental safety and economic considerations. There is a particular low solubility of **metal sulfides**, compared to the respective carbonate, phosphate, and oxide compounds. One major prerequisite is the microbial reduction of sulfate; thus, this process is particularly important in the marine environment, whereas in anoxic freshwaters milieu there is a tendency for enhancing metal mobility due to the formation of stable complexes with ligands from decomposing organic matter. Marine sulfidic conditions, in addition, seem to repressing the formation of mono-methyl mercury, one of the most toxic substances in the aquatic environment, by a process of disproportionation into volatile dimethyl-mercury and highly insoluble mercury-sulfide (Craig & Moreton, 1984). There are indications, that degradation of highly toxic **chlorinated hydrocarbons** is enhanced under in the sulfidic environment relative to oxic conditions (Sahm *et al.*, 1986). A summary of the positive and negative effects of anoxic conditions on the mobility of heavy metals, arsenic, methyl mercury and organochlorine compounds in dredged sludges is given in Table 7-4 (after Kersten, 1988).

Table 7-4 Summary of Positive and Negative Effects of Anoxic Sulfidic Conditions in Sludges (after Kersten, 1988).

Element or Compound	Advantageous Effects	Disadvantageous Effects
Heavy Metals (e.g.cadmium)	Sulfide precipitation	Formation of mobile polysulfide and organic complexes under certain conditions with low $Fe(OH)_3$ concentrations; strong increase mobility under acidic conditions
Metalloids (e.g.arsenic)	Capture by sulfides	Highly mobile under post-oxic & neutral/slightly alkaline cond.
Methyl mercury	Degradation/inhibition of CH_3Hg^+-formation by precipitation of HgS	Formation of mobile polysulfide complexes, especially at low Fe concentrations
Organochlorine compounds	Initiation of biodegradation by reductive dechlorination (methanic environment is more favourable)	Formation of harmful terminal residues with certain compounds especially in sulfidic environments; mobilisation through colloidal matter in pore waters

Marine Disposal

While on the first look disposal in the sea seems to have its merits due to considerable dilution, it has to be stated that due to the "open" character of both dispersive processes and the food chain the effects are mostly unpredictable. Table 7-5 (from the "Definitive Environmental Impact Report (D-EIR) Disposal of Dredged Material" of the Netherlands Ministry of Public Health) lists several **criteria**, which are particularly important for the marine disposal option of dredged materials.

Table 7-5 The Nature of the Environmental Impact Produced by the Disposal of Dredged Material at Sea (Ministerie van Volksgezondheid en Milieuhygiene, The Netherlands, 1979)

ABIOTIC ENVIRONMENT	
Soil	Silt formation through sedimentation, contamination of the soil in the sedimentation area
Groundwater	Not applicable
Surface Water	Turbidity caused by suspended silt particles; dispersal of contaminants by sea currents
Air	No important effects
BIOTIC ENVIRONMENT	
Disposal Site	Interferences with existing ecosystems; changes in the soil structure resulting in changes in composition and species of flora and fauna; accumulation of contaminants in benthic organisms
Environs of Disposal Site	Dispersion of contaminants with suspended silt particles and plankton as a result of currents; accumulation in the food chain of fishes and bottom living organisms via plankton and bottom sediment
HUMAN ENVIRONMENT	
Foodstuffs	Possibility of disappearance of commercially important species of fish; contamination of fish or crustaceans
Recreation	Possible negative effects on beach formations

All dredged materials, whether or not contaminated, have a significant **physical impact** at the point of disposal, which includes covering of the seabed (and smothering of benthic organisms) and local enhancement of suspended solids levels. In certain circumstances disposal may interfere with migration of fish (e.g. the impact of high turbidity on salmonids) or of crustacea (e.g. if deposition occurred in the coastal migration path of crabs).

The Oslo Commission Guidelines for the disposal of dredged material into the sea regulate (13th Meeting of the Standing Advisory Committee for Scientific Advice, Amsterdam, 10 - 14 March 1986), among others, that for the dredged material to be disposed of at sea the following **information** should be obtained: (a) amount and composition, (b) amount of substances and materials to be deposited per day (per week/month), (c) form in which it is presented for dumping, i.e. whether as a solid, sludge or liquid, (d) Physical (especially solubility and specific gravity), chemical, biochemical (oxygen demand, nutrient production) and biological properties (presence of viruses, bacteria, yeasts, parasites, etc.), (e) toxicity, (f) persistence, (g) accumulation in biological materials or sediments, (h) chemical and physical changes of the waste after release, including possible formation of new compounds, and (i) probability of production of taints reducing marketability of resources (fish, shellfish, etc.). These guidelines include advice on dredged material **sampling and analysis**, e.g. as to suitable numbers of separate stations for a certain amount of dredged material, frequency of sampling, etc. Characteristics of dumping site and method of deposit have to be evaluated, e.g., location in relation to living resources and to amenity areas, initial dilution achieved, dispersal, horizontal transport and vertical mixing characteristics.

Sub-Sediment Deposition

In a review of various marine disposal options Kester *et al.* (1983) suggested that the best strategy for disposing contaminated sediments is to isolate them in a **permanently reducing** environment. Disposal in capped mound deposits above the prevailing sea-floor, disposal in subaqueous depressions, and **capping** deposits in depressions provide procedures for contaminated sediment (Bokuniewiscz, 1983; Morton, 1983); in some instances it may be worthwhile to excavate a depression for the disposal site of contaminated sediment than can be capped with clean sediment. This type of waste deposition under stable anoxic conditions, where large masses of polluted materials are covered with inert sediment became known as "subsediment-deposit"; the first example was planned for highly contaminated sludges from Stamford Harbour in the Central Long Island Sound following intensive discussions in the U.S. Congress (Morton, 1980).

7.4. Treatment of Strongly Contaminated Sludges

During the last few years several unit operations have been developed for the treatment of contaminated residues such as dredged materials. Van Gemert *et al.* (1988) give a review of new aproaches (Figure 7-2):

- "A" - **Large scale concentration techniques.** These techniques are characterized by large scale applicabilities, low costs per unit of residue to be treated and a low sensitivity to variations in circumstances. It is advantageous that these techniques may be constructed in mobile or transportable plants. "A"-techniques include methods such as hydrocyclonage, flotation, and high gradient magnetic separation.
- "B" - **Decontamination or concentration techniques** which are especially designed for relatively small scale operation. These techniques are generally suited for the treatment of residues which contain higher concentrations of contaminants; they involve higher operating costs per unit of residue to be treated; furthermore, they are more complicated, require specific experience of operators and are suitably constructed in stationary or semi-mobile plants. "B"-techniques include biological treatment, acid leaching of inorganic compounds, ion exchange methods and solvent extraction of organic compounds.

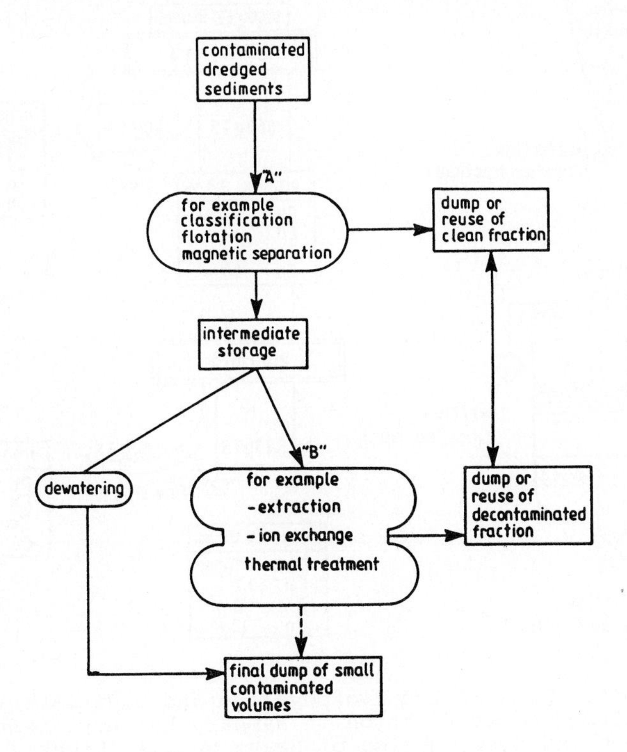

Figure 7-2 TNO-Scenario for the Treatment of Contaminated Dredged Sediments (Van Gemert *et al.* 1987).

As an example of "A"-techniques, the **classification of harbour sludge** from Hamburg with a highly effective combination of hydrocyclonage and elutriator as designed by Werther (1988) is shown in Figure 7-3. In the **hydrocyclone** the separation of the coarse fraction (relative clean sand) from the highly polluted fines is effected by the action of centrifugal forces. The coarse fraction leaved the cyclone in the underflow, while the fines are contained in the overflow. The advantage of the hydrocyclone is its simplicity and its ability to handle large throughputs; a disadvantage isthat the sharpness of the separation is fairly low. The **elutriator**, which follows in the classification scheme, effects a much better sharpness of separation. The basic principle is here separation according to the settling velocity of the particles in an upflowing water stream. There are virtually no fines found in the underflow of the elutriator.

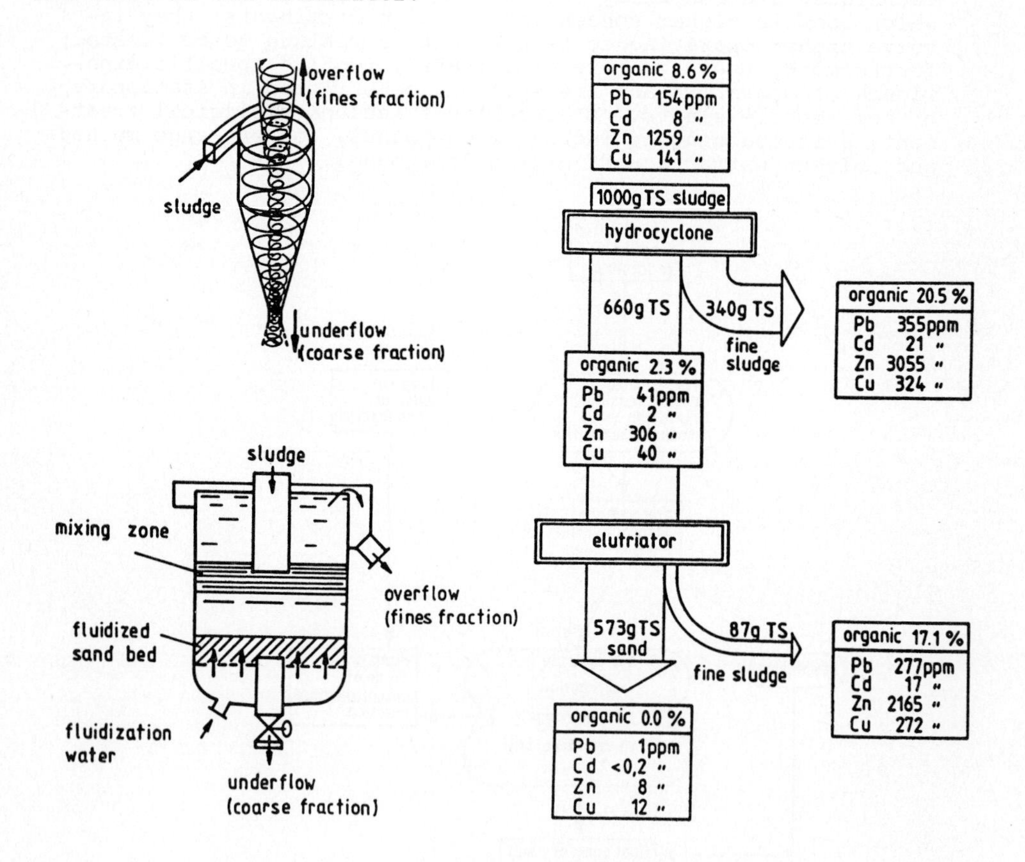

Figure 7-3 Schematic View of the Hydrocyclone and Elutriator Applied for Sludge Classification in Hamburg Harbour (Left). Mass Balance and Distribution of Heavy Metals (Right). After Hilligardt *et al.* (1986). TS = Dry Material.

The efficiency of the separation procedure is illustrated in Figure 7-3 (right) which demonstrates that the sand thus separated from the sludge has a heavy metal content which is of the same order of magnitude found in naturally occurring sandstones.

An example for "B"-techniques is the **acid leaching method** described by Müller & Riethmayer (1982), which comprises the following processing steps for extraction of metals from dredged materials:

(a) **Extraction** of metals with acid at pH = 0.5-10; extraction time 10-15 min;

(b) **Separation** of purified sludge from the acid solution followed by rewashing with decanting centrifuges;

(c) **Precipitation** of metals from the solution through pH increase with lime followed by carbonate precipitation with the CO_2 released with acid leaching. With this process extremely low residual concentrations of the metals are attained.

According to Van Gemert *et al.* (1988) the application potential of this method is determined by ecological and economic aspects. With respect to the former aspect, it is necessary to discharge waste water with large amount of salts; it is also uncertain, whether undesirable compounds are formed when organic matter is present in the sludge. Economic aspects include costs of alternative techniques or storage and the availability of a sufficient amount of HCl at a reasonable price and quality.

Stabilization Techniques

In general, solidification/stabilization technology is considered a last approach to the management of hazardous wastes. The aim of these techniques is a stronger **fixation** of contaminants to reduce the emission rate to the biosphere and to retard exchange processes. Two objectives can be distinguished:

(i) At best, a material is produced which can be used for land scape modification and constructions, e.g. of road, dikes, and walls for noise retardation; in this case disposal can be avoided;

(ii) in other cases the material could be improved by a stabilization process in order to deposit it safely and/or cheaply;

Most of the stabilization techniques aimed for the immobilization of **metal-containing wastes** are based on additions of cement, water glass (alkali silicate), coal fly ash, lime or gypsum. Detailed presentations and discussions of stabilization techniques of hazardous wastes and remedial actions for redevelopment of contaminated soils is given by Malone *et al.* (1982) and Wiedemann (1982). A review on stabilization of dredged mud is presented by Calmano (1988).

Laboratory studies on the evaluation and **efficiency** of stabilization processes were performed by Calmano *et al.* (1986). As an example Figure 7-4 shows acid titration curves for Hamburg harbor mud without and after addition of limestone and cement/fly ash stabilizers. Best results are attained with calcium carbonate, since the pH-conditions are not changed significantly upon addition of $CaCO_3$. Generally, maintainance of a pH of neutrality or slightly beyond favours adsorption or precipitation of soluble metals (Gambrell *et al.*, 1983). On the other hand it can be expected that both low and high pH-values will have unfavourable effects on the mobility of heavy metals.

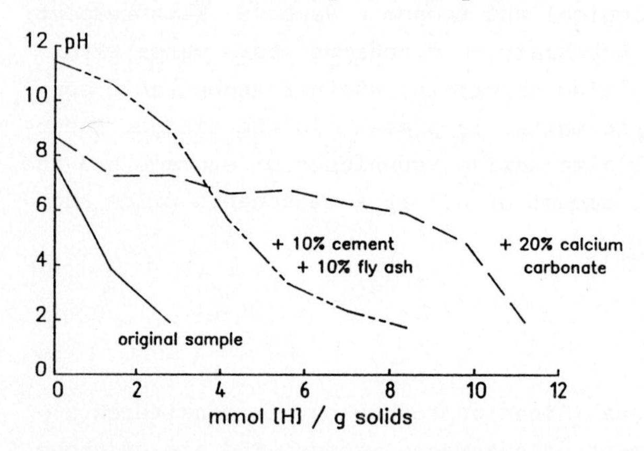

Figure 7-4

Effect of calcium carbonate and cement/fly ash additives on chemical stabilization of fine-grained sediment from Hamburg harbour (Calmano et al., 1986)

Regarding the **various containment strategies** is has been argued that upland containment, e.g. on heap-like deposits, could provide a more controlled management than containment in the marine environment. However, contaminants released either gradually from an imperfect impermeable barrier (also to groundwater) or catastrophically from failure of the barrier could produce substantial damage (Kester *et al.*, 1983). On the other hand, near-shore marine containment, e.g. in capped mound deposits, offers several advantages, particularly with respect to the protection of groundwater resources, since the underlying water is saline and chemical processes are favourable for the immobilisation or degradation of priority pollutants.

7.5. Case Study: Dredged Material from Rotterdam Harbor

The Netherlands have problems with dredged material since long (Salomons *et al.*, 1982). In the **Rhine River flood plain** contaminated sediments both affect groundwater quality and agricultural products:

> Calculations by Kerdijk (1981) of **dispersion processes** indicated that chloride, showing conservative behaviour, will appear in the adjacent polders in approx. the year 2100, the heavy metals 1 - 3 centuries, and pesticides several thousand years later.
>
> The **cattle grazing** in the river flood plains is exposed to three sources of heavy metals in the diet; the drinking (river) water, the herbage and the ingested soil particles. Contribution of **drinking water** is less than 1% of the total heavy metal intake. The contribution of the contaminated grass and soil particles, however, is quite significant for the daily intake of As, Cd, Hg, and Pb by a dairy cow on the river flood plain as compared with an animal in an uncontaminated situation.
>
> Pot experiments and field studies on dredged materials from Rotterdam harbour that in particular cadmium concentrations approximate or exceed the tolerable level in several **crops**. In green leafy vegetables, wheat grain and other cereals, and in most green fodder crops, the cadmium concentrations must be considered too high for human and animal consumption.
>
> While the levels of **chlorinated hydrocarbons** in Rhine sediments are relatively low and do not induce intolerable high concentrations in the consumable products, sometimes, pesticides and by-products of the synthesis thereof have been dumped in certain harbour areas; dredged materials originating from these areas and in use for agriculture, gave rise to intolerable contents of chlorinated hydrocarbons in potatoes and carrots.

In the **Rotterdam harbour** area, both fluvial and marine sediments are deposited. In the most easterly habours the percentage of marine mud is low whereas in the westerly harbours (Europoort) up to 90% of the mud may be derived from the North Sea. This decrease in the amount of fluvial, contaminated material is reflected in decreasing heavy metals in the seaward direction. From these compositional and geographical differences 4 classes of dredged materials are distinguished, for which different disposal options are being applied (Nijssen, 1988).

With respect to the more problematic **dumping activities** it was decided by the Definitive Policy Plan of the Netherlands (1982) that of the dredged sludge from Rotterdam harbour only **class 1 material** from the western harbour area, which is primarily of marine origin, is permitted for **disposal in the sea** at the Loswal Noord site (approx. 13 million m^3 per year).

Since beginning of 1985, the Municipality of Rotterdam has not been given an exemption under the Sea Water Act to discharge dredged material from the class 2 area (Botlek area) into the sea. That the disposal of class one dredged material at sea is considered acceptable is based on present knowledge. However, in the Policy Plan it is recommended that further research into this is necessary, and it is also established that the formulation of target values for the quality of sea water and bottom should be developed, in relation to the disposal of dredged material. In this context, is seems particularly important to know the **routes** of fine-grained and more polluted components even of class 1 material which may be washed out from bulk sediment during dumping and may enter into the **circulation** of the southern North Sea (Salomons *et al.*, 1988).

For the **disposal** of 1,5 million m^3 heavily contaminated, **class 4**, dredged material from localized activities such as those occurring at shipyards or the discharge points for industrial effluents, a special site has been prepared in the outer estuary on the Maasvlakte. The site is prepared in such a way that the dredged material is placed 0.5 m above the highest groundwater level (a criterion from the Waste Material Act) and in order to protect the quality of the groundwater, the dredged material is packed in a liner. The effectiveness of the plastic sheet is monitored by measuring in levelling tubes and by a horizontal drainage system beneath the liner. The surplus water that is released on the upper surface after the disposal is discharged via a settling basin into the Nieuwe Waterway. In the settling basin provision is made to remove floating layers, e.g. films of oil resulting from the discharge of material containing oil (Nijssen, 1988).

The use of the class 4 dredged material site has made possible the removal of the **heavily contaminated** silt which is found in specific localities in the harbours and fairways of the estuary. In the first phase, sections of the Petroleum Harbour were dredged to the required depth and the material was deposited in previously excavated pits in the Botlek Harbour. Specific conditions were laid down to cover the methods used to dredge and to discharge the contaminated silt. One major progress was the development of a special **"diffuser"** to discharge the material; this device leads to a limitation of the turbidity around the discharge point, because the diffuser takes the dredged material to the bottom as a cohesive density flow.

For the disposal of approx. 10 million m^3 dredged sludge of the **classes 2 and 3** from the harbour area the Port of Rotterdam and the Netherlands Waterways Administration - after several years of intensive and costly planning - have started to construct a **"sludge island"** in the form of a peninsula as a containment for approx. 150 million m^3 of sediment. The deposit will consist of a 20 m deep hole; the excavated material will form a 18 m high, high-tide resistant ring wall, containing approx. 30 million m^3 sand. With an area of 300 hectares the net capacity is about 90 million m^3. Owing to consolidation during filling a considerable volume reduction will occur; thus, approx 150 m^3 of wet sludge will eventually be deposited in this structure (Anon., 1984).

The sludge is transported via **pipeline** over a distance of approx. 2 km from "Mississippi harbour", to which the circulation water is pumped back after passing a purification unit. While the supernatant water resulting from the consolidation will be cleaned as well in this plant, part of the aqueous solutions from the deposit will enter the bottom sediment together with mobilized pollutants. **Model calculations**, however, suggest that the concentration of most contaminants will not affect groundwater composition significantly; it is expected that pollutants discharged to the seafloor will have only minor effects on the surrounding ecosystems (Anon., 1984).

It seems, that due to the short distance between the source and deposition areas in the case of Rotterdam harbour the large-scale "island solution" (5 - 8 Dutch guilders per m^3 original sediment) is **economically** well competitive to the old inland sites (5 - 10 D. Fl.) and to the sea disposal of dredged materials class 1 (3 - 7 D.Fl)(Nijssen, 1988). However, it has definitively been stated that the Municipality of Rotterdam has no intention to create further large-scale sites after 2002 (the official date by which time the present site is expected to be filled), and measures have to be undertaken to **improve the quality of the sediments**, particularly from municipal and industrial dischargers in the Rhine River catchment area, to be acceptable for other uses, such as civil engineering construction work, in the ceramics industry or for agricultural purposes.

References to 7.1 "Introduction"

d'Angremond, K. *et al.*, (1978) *Assessment of certain European dredging practices and dredged material containment and reclamation methods*. U.S. Army Engineer Waterways Exp. Stn., Techn. Rep. D-78-58. Vicksburg/MS: U.S. Army Corps of Engineers.

Anonymus (!981) *A survey of world port practices in the ocean disposal of dredged material as related to the London Dumping Convention*. Report of the Ad Hoc Dredging Committee, Mr. A.J. Tozzoli, Chairman. Port Authority of New York and New Jersey, 38 pp. New York: International Association of Ports and Harbors

Brannon, J.M. *et al.* (1976) Distribution of heavy metals in marine and freshwater sediments. In Proc. Conf. *Dredging and its Environmental Effects*, Mobile/Ala., eds. P.A. Krenkel *et al.*, pp. 455-495. New York: Amer. Soc. Civil Engineers.

Lee, C.R., Engler, R.M. & Mahloch, J.L. (1976) *Land Application of Waste Materials from Dredging, Construction and Demolition Processes*. Misc. Paper D-76-5. Prepared for Office, Chief of Engineers, U.S. Army Washington. Vicksburg: U.S. Army Corps of Engineers Waterways Experiment Station.

Lee, C.R. & Peddicord, R.K. (1988) Decision-making framework for management of dredged material disposal. In *Environmental Management of Solid Waste - Dredged Material and Mine Tailings*, eds. W. Salomons & U. Förstner, pp. 324-371. Berlin: Springer-Verlag.

Lee, G.F. & Plumb, R.H. (1974) *Literature Review on Research Study for the Development of Dredged Material Disposal Criteria*. DMRP Rept. D-74-1, 145 p. Vicksburg: U.S. Army Corps of Engineers Waterways Experiment Station.

Salomons, W. & Eysink, W. (1981) Pathways of mud and particulate trace metals from rivers to the southeastern North Sea. In *Holocene Marine Sedimentation in the North Sea Basin*, eds. S.D. Nio *et al.* Spec. Publ. *Int. Assoc. Sedimentol.* 5: 429-450.

Salomons, W. & Förstner, U. (1980) Trace metal analysis on polluted sediments. II. Evaluation of environmental impact. *Environ. Technol. Lett.* 1, 506-517.

Van Driel, W., Kerdijk, H.N. & Salomons, W. (1984) Use and disposal of contaminated dredged material. *Land Water Intern.* 53, 13-18

References to 7.2 "Environmental Impact of Dredging Operations"

Allan, R.J. (1986) *The Role of Particulate Matter in the Fate of Contaminants in Aquatic Ecosystems*. National Water Research Institute, Scientific Series No. 142, 128 p. Burlington/Ontario: Canada Centre for Inland Waters.

Boyd, B. *et al.* (1972) *Disposal of Dredge Spoil, Problem Identification and Assessment and Research Program Development*. Techn. Rept. H-72-8. Vicksburg/MS: U.S. Army Corps of Engineers Waterways Experiment Station.

Burns, N.M. & Ross, C. (1972) Project Hypo - discussion of findings. In *Project Hypo, An Intensive Study of the Lake Erie Central Basin Hypolimnion and Related Surface Water Phenomenon*, eds. N.M. Burns & C. Ross, Techn. Report TS-05-71-208-24, pp. 120-126. Burlington: Canada Centre for Inland Waters

Clark, R.R., Chian, E.S.K. & Griffin, R.A. (1979) Degradation of polychlorinated biphenyls by mixed microbial cultures. *Appl. Environ. Microbiol.* 37, 680-685.

Darby, D.A., Adams, D.D. & Nivens, W.T. (1986) Early sediment changes and element mobilization in a man-made estuarine marsh. In *Sediment and Water Interactions*, ed. P.G. Sly, pp. 343-351, New York: Springer-Verlag

Gambrell, R.P. *et al.* (1977) *Transformation of Heavy Metals and Plant Nutrient in Dredged Sediments as Affected by Oxidation and Reduction Potential and pH.* Dredged Material Res. Program, Rep. D-77-4, 309 p. Vicksburg/MS: U.S. Army Corps of Engineers.

Hebert, P. & Schwartz, S. (1983) *Great Lakes Dredging in an Ecosystem Perspective - Lake Erie*. Report Submitted to the Dredging Subcommittee of the Water Quality Board, 73 p. Ottawa: Environment Canada.

Khalid, R.A. (1980) Chemical mobility of cadmium in sediment-water systems. In *Cadmium in the Environment*, ed. J.O. Nriagu, Vol. 1: *Ecological Cycling*, pp. 257-304. New York: Wiley

Otsuka, H., Furuta, M. & Arakawa, Y (1976) Effect of dredging on heavy metal distribution in lake sediments. *Aichi-ken Kogai Chosa Senta Shoho* 1976, 57-63.

Saxena, J. & Howard, P.H. (1977) Environmental transformation of alkylated and inorganic forms of certain metals. *Adv. Appl. Microbiol.* 21, 185-226.

Seeyle, J.G., Hesselberg, G.J. & Mac M.J. (1982) Accumulation by fish of contaminants released from dredged sediments. *Environ. Sci. Technol.* 16, 459-464.

Sly, P.G. (1977) Some influences of dredging in the Great Lakes. In Proc. Intern. Symp. on *Interactions between Sediment and Freshwater*, ed. H.L. Golterman, pp. 435-443. The Hague & Wageningen: Junk & Pudoc Publ.

Sweeney, R. *et al.* (1975) Impacts of the deposition of dredged spoils on Lake Erie sediment quality and associated biota. *J. Great Lakes Res.* 1, 162-170.

Weber, W.J., Posner, J.C. & Snitz, F.L. (1982) *Water Quality Impacts of Dredging Operations*. U.S. EPA Grant No. R-801112. Grosse Ile/-Mich.: Grosse Ile Laboratory

Windom, H.L. (1975) Heavy metal fluxes through salt-marsh estuaries. In *Estuarine Research*, ed. L.E. Cronin, Vol. 1, pp. 137-152. New York: Academic Press

Windom, H.L. (1976) Environmental aspects of dredging in the coastal zone. *CRC Crit. Rev. Environ. Control* 5, 91-109

References to 7.3 "Disposal of Dredged Materials"

Anonymus (1979) *Definitief Milieueffectrapport Berging Baggerspecie.* The Hague/The Netherlands: Ministerie van Volksgezondheid en Milieuhygiene.

Anonymus (1986) *Oslo Commission Guidelines for the Disposal of Dredged Material.* Standing Advisory Committee for Scientific Advice for the Oslo Commission. Thirteenth Meeting, Amsterdam, March 10-14, 1986.

Bokuniewicz, H.J. (1983) Submarine borrow pits as containment sites for dredged sediments. In: *Dredged-Material Disposal in the Ocean* (Waste in the Ocean, Vol. 2), eds. D.R. Kester *et al.*, pp. 215-227. New York: Wiley

Brannon, J.M., Hoeppel, R.E. & Gunnison, D. (1984) Efficiency of capping contaminated dredged material. In: *Dredging and Dredged Material Disposal*, Vol 2. Proc. Conf. *Dredging '84*, pp. 664-673. Clearwater Beach/FL: Amer. Soc. Civ. Eng.

Cottenie, A. (1981) *Trace Metals in Agriculture and in the Environment.* Bruxelles: Instituut tot Aanmoediging van het Wetenschappelijk Onderzoek in Nijverheid en Landbouw (I.W.O.N.L.)

Craig, P.J. & Moreton, P.A. (1984) The role of sulphide in the formation of dimethylmercury in river and estuary sediments. *Mar. Poll. Bull.* 15, 406-408.

Folsom, B.L. jr., Lee, C.R. & Bates, D.J. (1981) *Influence of Disposal Environment on Availability and Plant Uptake of Heavy Metals in Dredged Material.* Techn. Rept. EL-81-12. Vicksburg/MS.: U.S. Army Corps of Engineers Waterways Experiment Station

Förstner, U. *et al.* (1986) Mobility of pollutants in dredged materials - implications for selecting disposal options. In *Role of the Ocean as a Waste Disposal Option*, ed. G. Kullenberg, pp. 597-615. Dordrecht: D. Reidel Publ. Co.

Gambrell, R.P., Khalid, R.A. & Patrick, W.H. jr. (1978) *Disposal alternatives for contaminated dredged material as a management tool to minimize environmental effects.* U.S. Army Engineers Waterways Experiment Stn., Techn. Rep. DS-78-8, Vicksburg/MS: U.S. Army Corps of Engineers

Herms, U. & Tent, L. (1982) Schwermetallgehalte im Hafenschlick sowie in landwirtschaftlich genutzten Hafenschlick-Spülfeldern im Raum Hamburg. *Geol. Jb.* F12, 3-11.

Kersten, M. (1988) Geochemistry of priority pollutants in anoxic sludges: cadmium, arsenic, methyl mercury, and chlorinated organics. In *Environmental Management of Solid Waste - Dredged Material and Mine Tailings*, eds. W. Salomons & U. Förstner, pp. 170-213. Berlin: Springer-Verlag.

Kester, D.R. *et al.* (Eds.)(1983) *Wastes in the Ocean*, Vol. 2: *Dredged-Material Disposal in the Ocean*, 299 pp. New York: Wiley

Lee, C.R. & Peddicord, R.K. (1988) Decision-making framework for management of dredged material disposal. In *Environmental Management of Solid Waste - Dredged Material and Mine Tailings*, eds. W. Salomons & U. Förstner, pp. 324-371. Berlin: Springer-Verlag

Maaß, B., Miehlich, G. & Gröngröft, A. (1985) Untersuchungen zur Grundwassergefährdung durch Hafenschlick-Spülfelder. II. Inhaltsstoffe in Spülfeldsedimenten und Porenwässern. *Mitt. Dtsch. Bodenkundl. Ges.* 43/I, 253-258.

Morton, R.W. (1980) "Capping" procedures as an alternative technique to isolate contaminated dredged material in the marine environment. In: *Dredge Spoil Disposal and PCB Contamination: Hearings before the Committee on Merchant Marine and Fisheries.* House of Representatives, Ninety-sixth Congress, 2nd Session, on Exploring the Various Aspects of Dumping of Dredged Spoil Material in the Ocean and the PCB Contamination Issue, March 14, May 21, 1980. USGPO Ser. No. 96-43, pp. 623-652, Washington DC.

Morton, R.W. (1983) Precision bathymetric study of dredged-material capping experiment in Long Island Sound. In: *Wastes in the Ocean*, Vol. 2: *Dredged-Material Disposal in the Ocean*, eds. D.R. Kester *et al.*, pp. 99-121. New York, Wiley.

Sahm, H., Brunner, M. & Schobert, S.M. (1986) Anaerobic degradation of halogenated aromatic compounds. *Microbial Ecol.* 12, 147-153

Salomons, W. *et al.* (1987) Sediments as a source of contaminants? In *Ecological Effects of* In Situ *Sediment Contaminants*, eds. R.L. Thomas *et al.*, *Hydrobiologia* 149, 13-30.

Van Driel, W. & Nijssen, J.P.J. (1988) Development of dredged material disposal site: Implications for soil, flora and food quality. In *Chemistry and Biology of Solid Waste - Dredged Material and Mine Tailings*, eds. W. Salomons & U. Förstner, pp. 101-126. Berlin: Springer-Verlag.

Van Driel, W., Smilde, K.W. & Van Luit, B. (1985) *Comparison of the Heavy Metal Uptake of Cyperus Esculentus and of Agronomic Plants Grown on Contaminated Dutch Sediments.* Misc. Paper D-83-1. Vicksburg/MS.: U.S. Army Corps of Engineers Waterways Experiment Station.

References to 7.4 "Treatment of Strongly Contaminated Sludges"

Calmano, W. (1988) Stabilization of dredged mud. In Environmental *Management of Solid Waste - Dredged Material and Mine Tailings*, eds. W. Salomons & U. Förstner, pp. 80-98. Berlin: Springer-Verlag

Calmano, W. *et al.* (1986) Behaviour of dredged mud after stabilization with different additives. In *Contaminated Soil*, eds. J.W. Assink & W.J. Van Den Brink, pp. 737-746. Dordrecht: Martinus Nijhoff Publ.

Gambrell, R.P., Reddy, C.N. & Khalid, R.A. (1983) Characterization of trace and toxic materials in sediments of a lake being restored. *J. Water Pollut. Control Fed.* 55, 1271-1279.

Hilligardt, R. *et al.* (1986) Classification and dewatering of dredged river sediments contaminated with heavy metals. 4th *World Filtration Congr.*, April 22-25, 1986. Ostend/Belgium.

Kester, D.R. *et al.* (Eds.)(1983) *Wastes in the Ocean*, Vol 2: *Dredged-Material Disposal in the Ocean*, 299 p. New York: Wiley

Malone, P.G., Jones, L.W. & Larson, R.J. (1982) *Guide to the Disposal of Chemically Stabilized and Solidified Waste*. Rept. SW-872, Office of Water and Waste Management. Washington D.C.: U.S. Environmental Protection Agency.

Müller, G. & Riethmayer, S. (1982) Chemische Entgiftung: das alternative Konzept zur problemlosen und endgültigen Entsorgung schwermetallbelasteter Baggerschlämme. *Chem. Ztg.* 106, 289-292.

Van Gemert, W.J.Th., Quakernaat, J. & Van Veen, H.J. (1988) Methods for the treatment of contaminated dredged sediments. In *Environmental Management of Solid Waste - Dredged Material and Mine Tailings*, eds. W. Salomons & U. Förstner, pp. 44-64. Berlin: Springer-Verlag.

Werther, J. (1988) Classification and dewatering of sludges. In *Environmental Management of Solid Waste - Dredged Material and Mine Tailings*, eds. W. Salomons & U. Förstner, pp. 65-79. Berlin: Springer-Verlag.

Wiedemann, H.U. (1982) *Verfahren zur Verfestigung von Sonderabfällen und Stabilisierung von verunreinigten Böden*. Berichte Umweltbundesamt 1/82. Berlin: Erich Schmidt Verlag

References to 7.5 "Case Study: Dredged Material from Rotterdam Harbor"

Anonymus (1984) *Grootschalige locatie voor de berging van baggerspecie uit het benedenrivierengebied.* Projectreport/Environmental Compatibility Study, Oct. 1984, 334 p. Rotterdam: Municipality of Rotterdam/Rijkswaterstaat

Kerdijk, H.N. (1981) Groundwater pollution by heavy metals and pesticides from a dredge spoil dump. In *Quality of Groundwater*, eds. W. Van Fuyvenboden, P. Glasbergen & H. Van Lelyveld, pp. 279-286. Amsterdam: Elsevier Publ. Co.

Nijssen, J.P.J. (1988) Rotterdam dredged material: Approach to handling. In *Environmental Management of Solid Waste - Dredged Material and Mine Tailings*, eds. W. Salomons & U. Förstner, pp. 243-281. Berlin: Springer-Verlag.

Salomons, W. *et al.* (1982) Help! Holland is plated by the Rhine (environmental problems associated with contaminated sediments). *Effects of Waste Disposal on Groundwater. Proc. Exeter Symp. IHAS* 139, 255-269.

SUBJECT INDEX